petroleum refinery distillation

r.n. watkins

petroleum refinery distillation

gulf publishing company
houston, texas

PETROLEUM REFINERY DISTILLATION

ISBN 0-87201-671-4

Library of Congress Catalog Card Number: 72-93693

contents

ac knowl edg ments

The author wishes to extend appreciation to the following individuals for their assistance in the development of this work. To B.P. Castiglioni and E.L. Ekholm of the Bechtel Corporation for their review of the original manuscript; to J.S. Gusler of J.F. Pritchard & Co. and D.E. Roper of Eastham and Meyer for their invaluable assistance in clearing certain portions for publication; to Harold Hoffman, Clayton Umbach, Phyllis Hall and Irene Karlik of Gulf Publishing Company for their assistance in preparing the manuscript for publication; to my typists, Mardella Van Eman, Pat Turner and Sherry Harvey; and to my wife, Lynne Watkins, who prepared all the art work for this book in addition to showing extreme patience and providing much-needed support during many phases of this project. Thanks is also extended to Creole Petroleum Corporation for their gracious permission to publish the material relative to the vacuum tower in Chapter 3.

R.N. Watkins
Sugar Land, Texas
January 2, 1973

petroleum refinery distillation

introduction

This book was written to fill a void which exists in the technical literature which the author encountered early and often in his career. Although petroleum distillation has been practiced for over a century, there has been precious little published literature in the field. In the field of articles, Packie's (1) classical work has been, up to this very day, the first and best basic. Later work by Prater and Boyd (2) and by Watkins (3) completes the existing short articles available. In textbooks, a brief treatment by Nelson (4) and by van Winkle (5) is all that one can locate as reference material. The situation is not much better in the private confidential literature.

Several years ago, I joined a major engineering-construction firm after having spent over six years in the petrochemical industry. At that time, my practical knowledge of petroleum processing was slight. As luck would have it, I was soon called upon to design a petroleum-processing tower. After receiving brief and very general guidelines, I got underway. My first approach was to seek help from more experienced men in the firm. At this point, I received the shock of my life. Since this particular tower was a little different from the conventional oil plant, no one could be very helpful because, while they might have worked previously on a refinery design, they were not particularly knowledgeable about the variables one must consider in petroleum distillation. And so, I turned to the literature—and found nothing. The library is full of books covering distillation—theory, practice, equilibrium, equipment design and every other conceivable topic. Trade journals continue to carry "new and improved" calculation techniques for conventional distillation. Why so little on petroleum fractionation? I suspect that the answer to this question lies in two areas. First, until Edmister's book (6) appeared in 1961, there were no easy-to-use correlations for vapor-liquid equilibrium in heavy oils. Secondly, many engineers today are more interested in studying computer solutions of engineering problems wherein the method of solving the problem is the primary concern followed at a discrete distance by the consideration of the chemical engineering fundamentals involved.

This book is intended to be, first of all, practical and usable by the working engineer and is primarily addressed to him. Secondarily, it is intended to present and, where needed, to clarify proven principles and methods relating to petroleum distillation. It is a fact that techniques in this area of technology and much of the available design data are largely empirical, but, considering that petroleum is a mixture of a very large number of individual components of many different types, how could one expect otherwise? While it would be highly desirable to be able to analyze petroleum distillation with the same degree of clarity and knowledge as one might the separation of ethyl alcohol and water, it is unreasonable to demand such approaches to these problems. Empiricism, while often an anathema to the academic scholar, does provide the man in the field with working solutions to his daily problems. Empirical methods have yet another asset and one not always true of theoretical explanations—they work—for they rationalize and attempt to explain observed phenomena which have resisted substantiation by theory.

As in any technical book being written today, there is little that can be claimed as radically new. Much of the new data and methods being made available today are in the form of articles in journals or papers delivered at symposia. What is new about this book is that it assembles in one place for the first time in the public literature a complete set of design/analytical procedures for every type of fractionation process to be found in a modern petroleum refinery or petrochemical plant. These procedures are those currently in use throughout the industry and have been proven by performance.

The book is organized into chapters, each one describing in detail a particular type of distillation process and the applicable design procedures. Illustrative material is provided where required to clarify the narrative material. As to the sources of data used in developing the calculations and design correlations, The American Petroleum Institute—the API—has published a data book (7) which contains its recommendations as to the best methods available in the public domain for estimating physical properties of petroleum and petroleum fractions. Nelson and Maxwell (8) are also valuable sources of such data. For heavy oil vapor-liquid equilibrium calculations, the methods of Packie and Edmister are in general use. In the example calculations, the thermal data was that of Johnson and Grayson (9).

Please note that in many places, the narrative material refers to the "owner." The "owner" is defined as the party ultimately responsible for the acceptance and operation of the facilities under study. Many basic decisions at the beginning of a design should be made by the person who is ultimately going to be responsible for the operation of the unit even though many routine decisions must still be made by the process designer in the course of the job.

It is assumed by the author that the reader has a basic understanding of distillation, of crude petroleum of petroleum products and their specifications and of petroleum refinery processes. If this is not the case, he is referred specifically to Nelson and to Hengstebeck (10) for background information.

As a final word of explanation and caution, the methods presented in this book are not intended as *the one, only, sacred way to do it.* These procedures have been developed by the author as logical approaches to solving problems and represent one man's experience and opinions. The reader is encouraged to examine these methods critically in the light of his own experiences and to govern himself accordingly.

References

1. J.W. Packie, "Distillation Equipment in the Oil Refining Industry," *AIChE Transactions* 37 (1941), pp. 51-78.
2. N.H. Prater and C.W. Boyd, "How to Calculate Multidraw Towers," *Oil and Gas Journal* (May 2, 1955), pp. 72-82.
3. R.N. Watkins, "How to Design Crude Distillation," *Hydrocarbon Processing* 48, no. 12 (December, 1969), pp. 93-106.
4. W.L. Nelson, *Petroleum Refinery Engineering*, 4th ed. (New York: McGraw-Hill Book Company, Inc., 1958).
5. M. van Winkle, *Distillation* (New York: McGraw-Hill Book Company, Inc., 1968).
6. W.C. Edmister, *Applied Hydrocarbon Thermodynamics*, (Houston: Gulf Publishing Company, 1961).
7. *Technical Data Book—Petroleum Refining* (Washington, D.C.: The American Petroleum Institute, 1966).
8. J.B. Maxwell, *Data Book on Hydrocarbons*, (Princeton, N.J.: D. van Nostrand Co. 1965).
9. Grayson and Johnson, *Petroleum Refiner*, 42, no. 2 (February, 1961), p. 123.
10. R.J. Hengstebeck, *Petroleum Processing, Principles and Applications* (New York: McGraw-Hill Book Company, Inc. 1959).

atmospheric tower

Crude petroleum as it is produced from the field is a relatively low-value material since, in its native state, it is rarely usable directly. However, it can be refined and further processed into any number of products whose value is many times that of the original oil. The first step in any petroleum refinery is the separation of the crude into various fractions by the process of distillation. These fractions may be products in their own right or may be feedstocks for other refining or processing units.

In most refineries, this process is carried out in two stages. The oil is first heated to the maximum temperature allowable for the crude being processed and for the operation being practiced and then fed to a fractionating tower which operates at slightly above atmospheric pressure. This tower is usually called the *atmospheric tower*. It yields several distillate products and a bottoms product which is the residual liquid material which could not be vaporized under the conditions of temperature and pressure existing in the atmospheric tower. This bottoms liquid is then reheated to the maximum allowable temperature–usually higher than the maximum temperature allowed for the feed to the atmospheric tower–and fed to a fractionating tower which operates at subatmospheric pressure. This tower is usually called the *vacuum tower*. This chapter describes the atmospheric tower. The vacuum tower is discussed in Chapter 3.

Although the process of crude oil distillation has been practiced in the United States for more than 100 years, the design and operation of these units is still accomplished almost exclusively on an empirical and, in some cases, even a nearly emotional basis. This is undoubtedly due to the fact that crude oil is made up of an almost infinite number of discrete hydrocarbons, all the way from methane to materials having 70 or more carbon atoms.Thus, the process and/or its products are usually discussed in terminology which is unique to the petroleum industry. In this introductory material, the nomenclature usually employed in crude oil distillation will be discussed before proceeding into the design methods.

Analysis of Crude Petroleum and Its Fractions

A complete component-by-component analysis of a crude oil sample is not practically realizable. For this reason, the composition of any given oil is approximated by a *true boiling point* distillation, commonly called TBP distillation. A thorough discussion of the equipment and procedures involved is given in Chapter 4 of Nelson (1). This method is basically a batch distillation using a large number of stages and a high reflux-to-distillate ratio so that the temperature at any point on the temperature-volumetric yield curve represents the actual (true) boiling point of the hydrocarbon material present at the volume percentage point.

TBP distillations are normally run only on crude oils and not on petroleum fractions. Instead, a rapid distillation procedure is used for analysis of petroleum products and

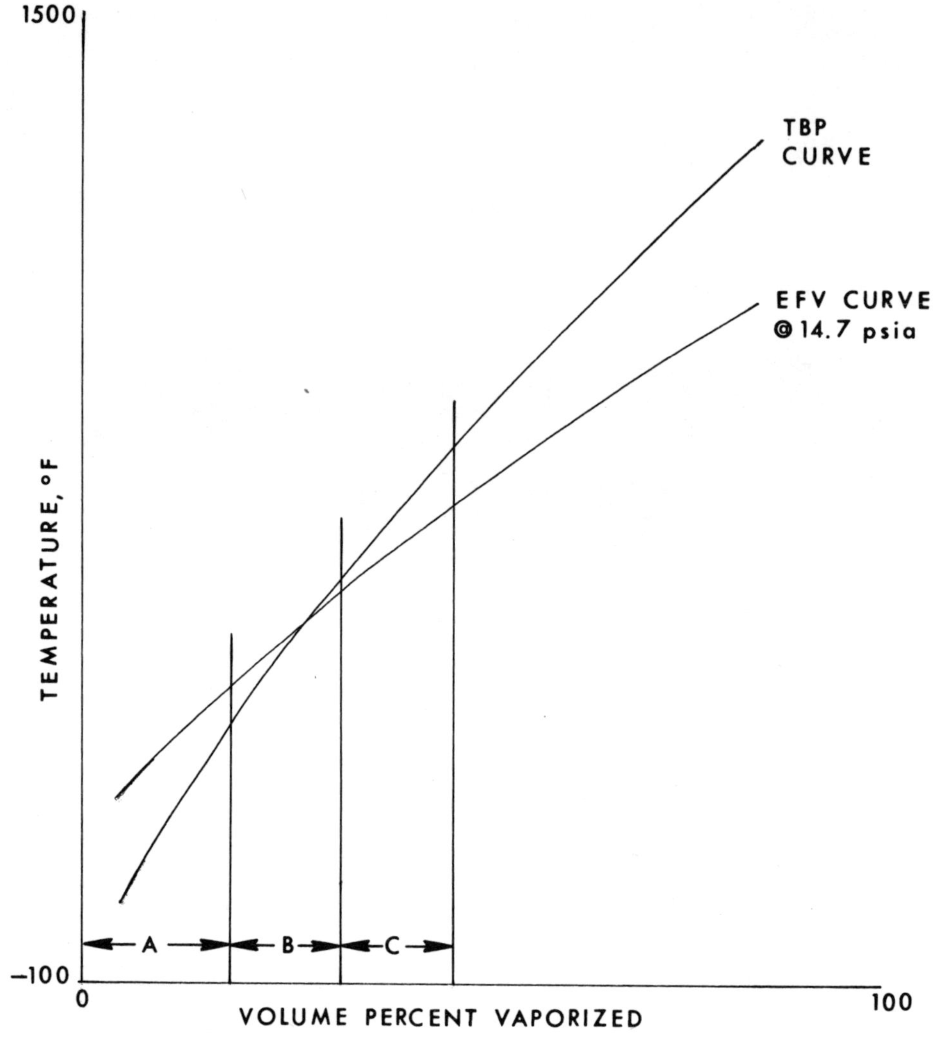

Figure 2.1. TBP and EFV Curves for a typical crude oil and product yield by volume.

intermediate fractions. These procedures were developed by the American Society for Testing Materials and are called ASTM distillations. These methods are rapid batch distillations employing no trays or reflux between the stillpot and the condenser. The only reflux available is that generated by heat losses from the apparatus. These test methods are used in control laboratories throughout the world.

The third type of laboratory distillation which is used in discussing petroleum fractionation is the *equilibrium flash vaporization* (distillation) commonly called the EFV. The procedure is also discussed in Chapter 4 of Nelson. This distillation can be run at pressures above atmospheric as well as under vacuum, whereas the TBP and ASTM distillations are run either at atmospheric pressure or under vacuum. EFV curves are seldom run because of the time and expense involved and are almost exclusively limited either to crude oil or to reduced crude samples (atmospheric tower bottoms liquid) which are being evaluated as vacuum tower charge stocks. The EFV initial boiling point is the bubble point of the fraction under study, and the EFV final boiling point is its dew point. If desired, a family of EFV curves can be run at a series of pressures encompassing the envisioned operating pressure. This will allow a more accurate estimate of phase behavior than can be attained from calculations.

The purpose for running these various distillations is to characterize the composition of the material under study, and, for this reason, the boiling range of petroleum liquids is of primary importance. For example, kerosene will have an ASTM boiling range of approximately 325 to 525 degrees F which corresponds to a TBP boiling range on the order of 275 to 575 degrees F. While the latter is theoretically indicative of actual composition, the former is the one used by both designers and operators in discussing distilla-

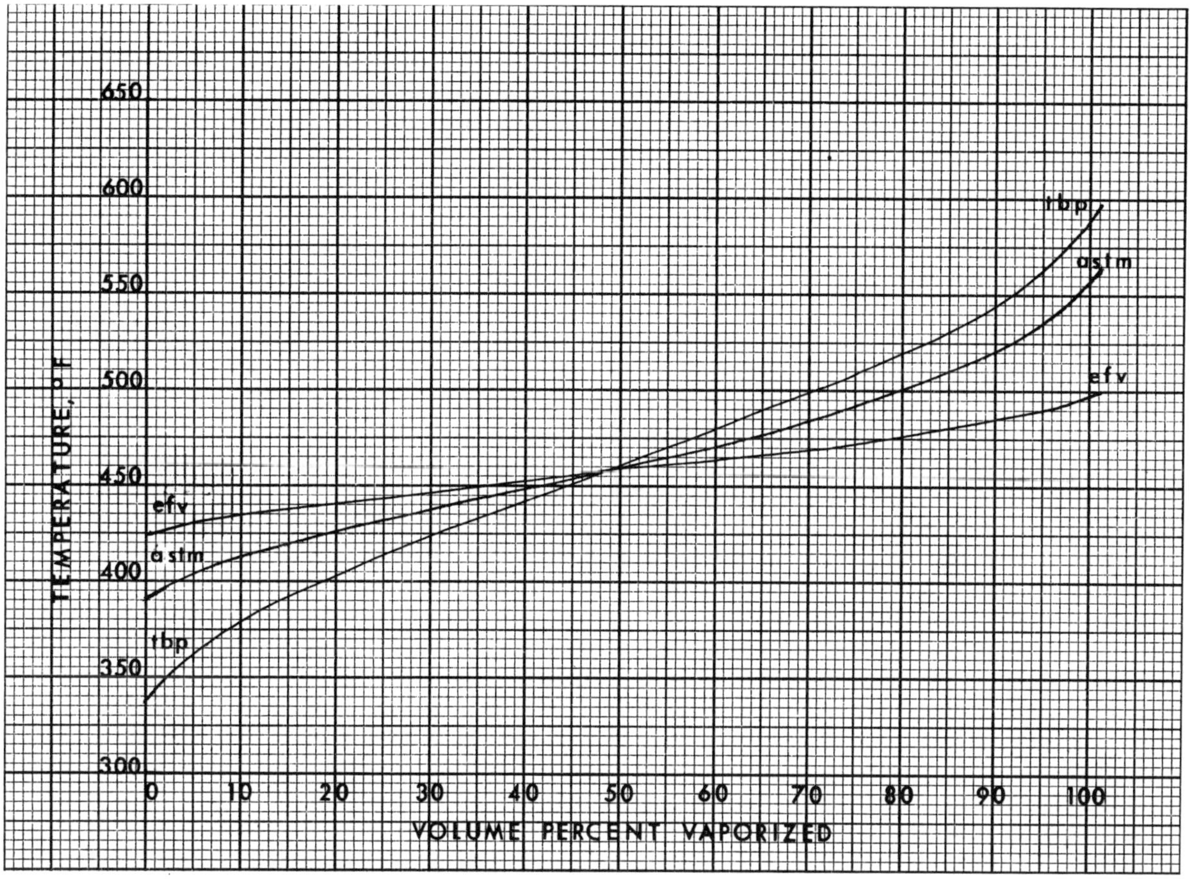

Figure 2.2. ASTM, TBP and EFV curves for 39.7 degree API light distillate.

tion problems. The EFV curve is used for determining phase behavior of liquids and, thus, is usually of interest principally to designers.

Figure 2.1 shows the TBP and atmospheric pressure EFV curves for a typical crude oil. Also illustrated is the planned yield of three distillate products for a process under study. Figure 2.2 shows the ASTM, TBP and atmospheric EFV distillation curves for a typical petroleum fraction.

The complete and definitive analysis of a crude oil, usually called a *crude assay*, is considerably more detailed than a TBP curve and a whole crude API gravity. A complete crude assay will contain some or all of the following:

1. Whole crude gravity, viscosity, sulfur content, pour point, etc.
2. TBP curve, mid-volume plot of gravity, viscosity, sulfur, etc.
3. Light-ends analysis up through C_8 or C_9.
4. Properties of fractions (naphthas, middle distillates, gas oils and residua)–yield as volume percent, gravity, sulfur, viscosity, octane number, diesel index, flash and fire point, freeze point, smoke point, pour point, vapor pressure, etc.
5. Properties of lube distillates (only if the crude is suitable for the manufacture of lube basestocks).
6. Properties of asphalts (only if the residua have suitable characteristics for preparation of asphalts).
7. Detailed studies of fractions for various properties, e.g., octane number versus yield for naphthas or viscosity versus yield for lubestocks.
8. EFV curve run at atmospheric pressure and/or phase diagram, although this is rarely done.

Much of this information is of little use in design work but is of great importance to the refiner or to a crude oil purchaser.

A typical assay of a commercial crude oil is included in the Appendix. This data was developed experimentally and correlated by Humble Oil & Refining Company and is reproduced here with their permission.

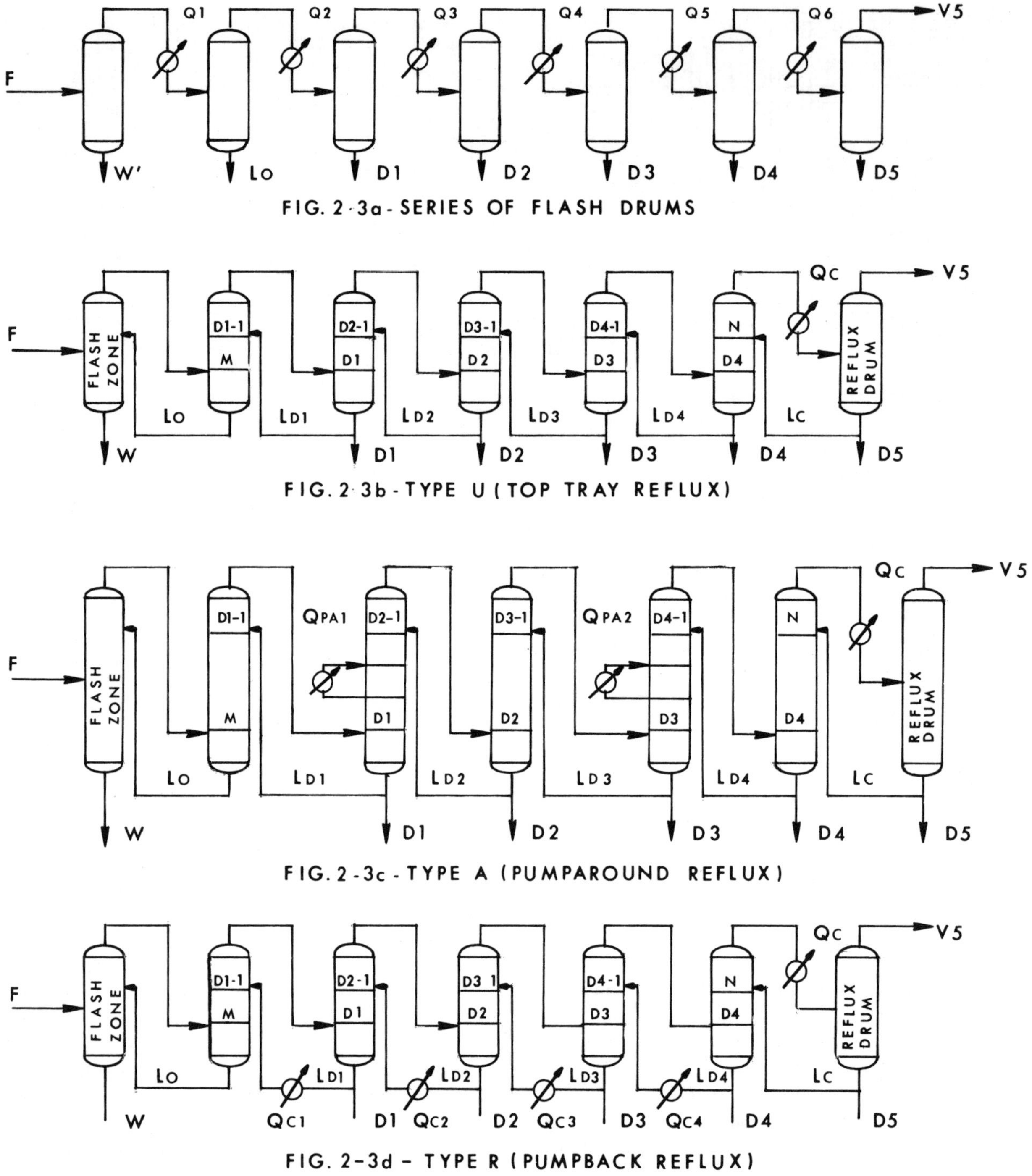

Figure 2.3. Processing schemes for atmospheric distillation of crude oil.

Basic Processes for Atmospheric Crude Distillation

The distillation of crude petroleum into fractions having different properties is similar to many more familiar types of fractionation processes although, on the surface, this may not appear to be so. In order to avoid details which are of only secondary importance at this point and to get down to the basics of the problem, the separation of crude oil into its desired fractions will be considered separately, disregarding the effect of sidestream product and bottoms liquid stripping, either by steam or by reboiling in the case of the lighter sidestreams. Further, the separation will be defined as five distillate fractions and a residual (bottoms) liquid. The lightest distillate fraction is not always completely condensible at the conditions of temperature and pressure in the reflux drum and, thus, may be yielded as two distillate fractions, one vapor and one liquid. A small amount of extra vaporization, called overflash, will be employed to provide adequate reflux in the section between the flash zone, i.e., the point in the tower where the partially vaporized feed enters, and the first (lowest) sidestream product draw tray.

Because of the continuously increasing temperature of crude oil as more and more of the material is vaporized and because the charge is usually heated to its maximum allowable temperature prior to entering the tower, it is not feasible to use a reboiler in crude distillation. Thus, all distillate products plus the overflash must be vaporized up from the flash zone. Another way of stating the same principle is that the total heat input required by the process must be contained in the feed as it leaves the crude furnace.

Figure 2.3 illustrates various processes for yielding distillates from the feed. The nomenclature used in designating the three methods of refluxing the process is the author's own for the purposes of this discussion and has no significance in the petroleum industry. By way of explanation, Type U employs only top tray reflux from an overhead condenser for total process heat removal. Type A employs an externally circulated and cooled stream, called *pumparound reflux*, at one or more side points on the tower for partial process heat removal. Type R employs subcooling of true equilibrium reflux, called *pumpback reflux*, for partial process heat removal, this usually occurring at all sidestream product draw points. In both Type A and Type R, only a portion of the total heat removal can be accomplished by these techniques; the remaining portion of the removal must be satisfied by reflux from the condenser to the top tray.

For the sake of simplicity in discussing Figure 2.3, it is stipulated that the volumetric yield of each product from the crude feed will be the same, regardless of the method used for heat removal.

Consider Figure 2.3*a* as the first method for yielding the desired product pattern. The vapor-liquid mixture from the furnace enters the first flash drum where the residual liquid, W′, drops out. The first drum effluent vapor is cooled just enough to condense the overflash, L_0, and then enters the second drum where this vapor-liquid separation occurs. Although not shown on the sketch, L_0 combines with W′ to form the total residual liquid, W. The vapor is cooled again to some predetermined temperature and enters a third flash drum where D1 separates out. This process is repeated as D2, D3 and D4 are condensed and separated out. The overhead vapor from the sixth drum is cooled to as low a temperature as is possible consistent with the available cooling medium in order to maximize distillate liquid recovery in D5. The noncondensibles form the distillate vapor, V5. Each of these separations has occurred due to cooling and equilibrium condensation so that the liquid from each stage contains appreciable amounts of components lighter than the desired distillate. The equilibrium vapor also contains some of the components which should be in the liquid so that the next fraction to be condensed will have some undesirable heavy ends. All these fractions are portions of a continuum which indicates that it would be possible to use reflux to wash back the heavy components from the vapor and stripping vapors to remove light components from the sidestream products.

Figure 2.3*b* illustrates a Type U system, the first approach to genuine fractionation. Each one of the vessels in the flow line represents one separation section. This type of unit is obviously impractical since heat is removed only in the overhead condenser at a temperature level too low for recovery of heat. Secondarily, the vapor and liquid traffic throughout the column increases markedly from bottom to top, which requires sizing the tower based on flows existing at the top tray. As will be seen later, this will result in a much larger tower diameter than for either Type A or Type R unit. Figure 2.4*a* illustrates the flows around a Type U section. Inspection of the flow scheme reveals that this is merely a conventional rectifying column.

Heat removal at intermediate points on the tower can be attained by withdrawing an internal liquid stream from the tower, cooling it and returning it to the column. The cooling medium is usually the crude oil charge which is being preheated before entering the furnace. Thus, a dual benefit is realized. This can be done in one of two ways.

The more common of the two processes is by the use of pumparound reflux which the author calls a Type A system. This is illustrated by Figures 2.3*c* and 2.4*b*. Liquid is withdrawn from a tray above the lower draw, cooled and returned to a tray further up in the tower but below the upper draw tray. In normal practice, such a section generally uses two or three actual trays including the pumparound draw tray and the tray on which the cooled liquid reenters the tower. This technique is generally employed at a low and at an intermediate point in the tower and makes heat available for exchange against crude oil charge at two different temperature levels. This method has the advantage of stabilizing vapor and true reflux liquid flow throughout the section where it is used as well as significantly reducing vapor and liquid traffic throughout the column. The disadvantage of this method is that the three trays normally used for heat removal must be considered as only one actual tray for fractionation purposes. This is because that, from an equilibrium viewpoint, the pumparound liquid is foreign to the zone in which it is employed. This process is analagous to removing heat from an absorber by withdrawing part of

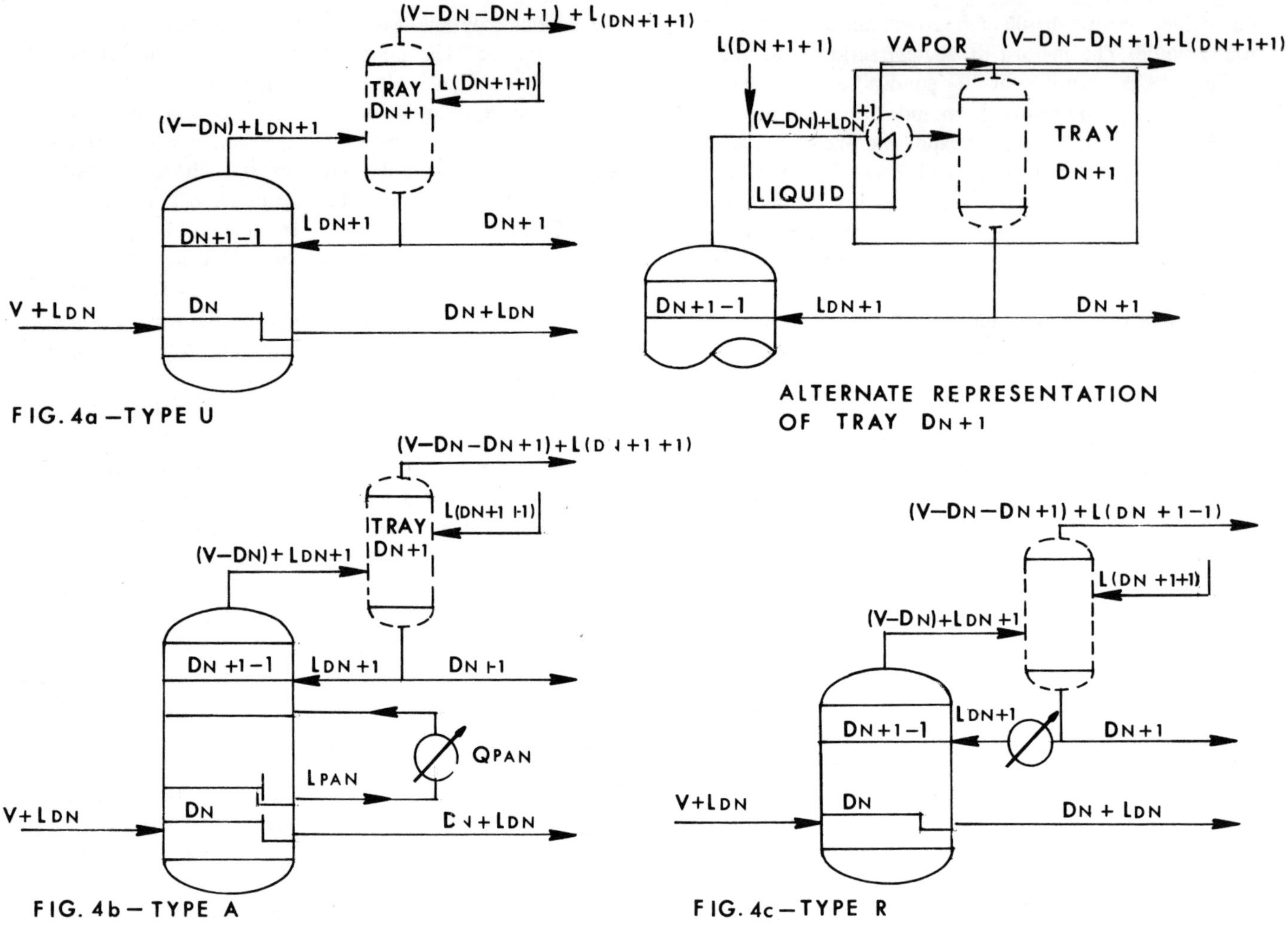

Figure 2.4. Detailed flows around a section for Types U, R and A atmospheric towers.

the bottoms, cooling it and returning it to an upper tray. This is fine for heat removal purposes but penalizes mass tranfer.

Figures 2.3*d* and 2.4*c* show a Type R system which is a rectifying section employing subcooled reflux. The most economical Type R design uses crude oil charge as the cooling medium for the pumpback reflux streams and, thus, also obtains the benefit of feed preheat. This is employed at all draw trays. There is the additional advantage that all liquids within the tower are true equilibrium reflux liquids, and, thus, all trays can be considered as being effective for fractionation purposes. A fairly uniform vapor and liquid traffic exists in all sections of the tower.

The typical vapor and liquid traffic which exists in the three types of refluxed columns is shown in Figure 2.5. Note that the Type U tower has continuously increasing vapor and liquid rates while the rates in both the Type A and Type R towers are more or less stabilized by the external heat removal.

Separation Criteria in Petroleum Fractionation

In the context that most chemical engineers apply to distillation of systems made up of discrete compounds identifiable by name, heavy oil fractionation is a black art, especially when encountered for the first time. Over the years, many excellent methods have been published for calculating all types of distillation systems involving discrete compounds. Even the inexperienced engineer is familiar with many of the basic techniques. Petroleum fractionation is quite a different story since the methods of analysis are largely empirical or are based on empirical criteria. However, there is a similarity between light hydrocarbon fractionation and crude oil distillation. It is the purpose of the following to demonstrate this similarity.

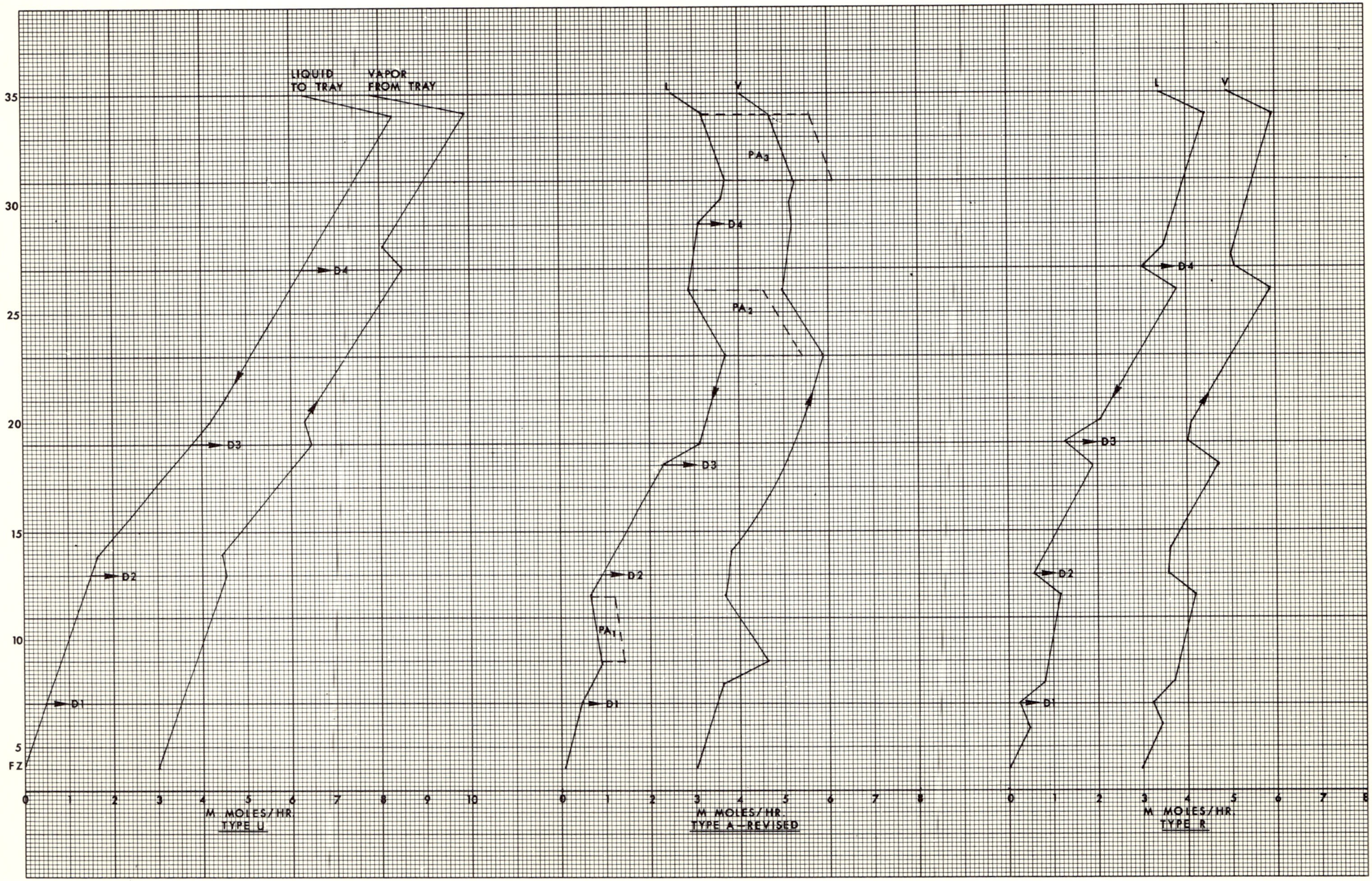

Figure 2.5. Vapor and liquid traffic for Types U, R and A atmospheric towers.

A typical light hydrocarbon separation is that of fractionating between propane and n-butane. This problem requires that the designer provide a sufficiency of trays and reflux to satisfy the composition specifications. Satisfying the trays-reflux requirements depends upon the specified separation requirements of the process and the inherent physical chemistry of the system involved. Talking in terms of a binary system, two terms which will be used later to clarify the language used in heavy oil work must be defined.

Degree of separation can be defined in terms of product purities or in terms of component recoveries. The greater the degree of separation, the greater will be the recovery of the light component in the distillate and the heavy component in the bottoms. This will result in higher product purities.

Degree of difficulty of separation is defined as the relative difficulty encountered in separating the two compounds in question, regardless of the purity requirements set by the process specifications. In light ends terminology, it may be considered as inversely proportional to the relative volatility between the two components.

From one's past experience, the following become obvious.

1. For a given system, the tray requirements increase markedly as purity requirements become greater, but reflux requirements increase only a small amount once a relatively high purity level is reached.
2. For a fixed separation, tray and reflux requirements increase as the relative volatility decreases, i.e., the separation becomes more difficult. For example, the propane and n-butane separation is easier than propane-propylene but more difficult than propane and n-pentane.

Speaking qualitatively, at reflux conditions exceeding minimum requirements, tray requirements are directly proportional to the required degree of separation and to the degree of difficulty of separation inherent in the physical-chemical system under consideration. Conversely, for a fixed number of trays, reflux requirements are directly proportional to the degree of difficulty inherent in the system and, to a somewhat lesser extent, to the required degree of separation.

The above terminology is the author's own. However, it is straightforward and, when applied to any distillation system involving discrete components, will result in a rapid qualitative assessment of tray-reflux requirements.

In the refinery, two terms are used to discuss product composition and the degree of separation between adjacent fractions. *ASTM boiling range* defines the general composition of the fraction and is usually one of the key specifications for most distillates from both the atmospheric tower and the vacuum tower.

The second term, *(5-95) Gap*, defines the relative degree of separation between adjacent fractions. It is determined by subtracting the 95 volume percent ASTM temperature of a fraction from the 5 volume percent ASTM temperature of the adjacent heavy fraction.

$$(5\text{-}95)\ \text{Gap} = (t_{5_H} - t_{95_L})_{ASTM},\ \text{degrees F}$$

Packie's (2) classic paper was the first to disclose criteria for defining fractionation between atmospheric tower distillate streams. Figure 2.6 is Packie's curve for fractionation between the overhead fraction and the adjacent sidestream.

The nomenclature for this correlation is as follows:

1. $L_{N'}$ = gallons per hour reflux from the top tray measured as 60 degree F liquid.
2. D_N = gallons per hour total distillates (vapor and liquid) to top tray, measured as 60 degree F liquid.
3. N_T = number of actual trays in the section, i.e., trays M through N inclusive = N – (M – 1) = N – M + 1. Note that each tray in pumparound heat removal service counts as one-third of an actual tray.

Figure 2.7 is Packie's curve for fractionation between sidestream products. The nomenclature for this correlation is as follows:

1. L_M = gallons per hour reflux from the upper draw tray measured as 60 degree F liquid.
2. P_N = gallons per hour total product vapors, measured as 60 degree F liquids, to the upper draw tray, i.e., stream D_N plus all lighter products.
3. N_T = number of actual trays in the section, i.e., M through (N – 1) inclusive = N – M. Note that each tray in pumparound heat-removal service counts as one-third of an actual tray.
4. $\Delta t(50\%)$ = (ASTM 50 percent temperature of the lower sidestream product, D_M) – (ASTM 50 percent temperature of the total products lighter than D_M).

Figure 2.7 does not apply to fractionation between the lowest sidestream and the bottoms stream, nor does it apply to vacuum fractionation although it is often used for the latter purpose due to lack of anything better. Note that these curves apply only for the case where steam stripping of sidestreams is practiced at rates of at least 0.2 pounds steam per gallon of product (8.4 pounds steam per barrel product). Reboiling of sidestreams will also satisfy this stipulation as long as the portion of the sidestream vaporized back is at least equal to that which would be produced by the above-mentioned steam rate. In cases where this

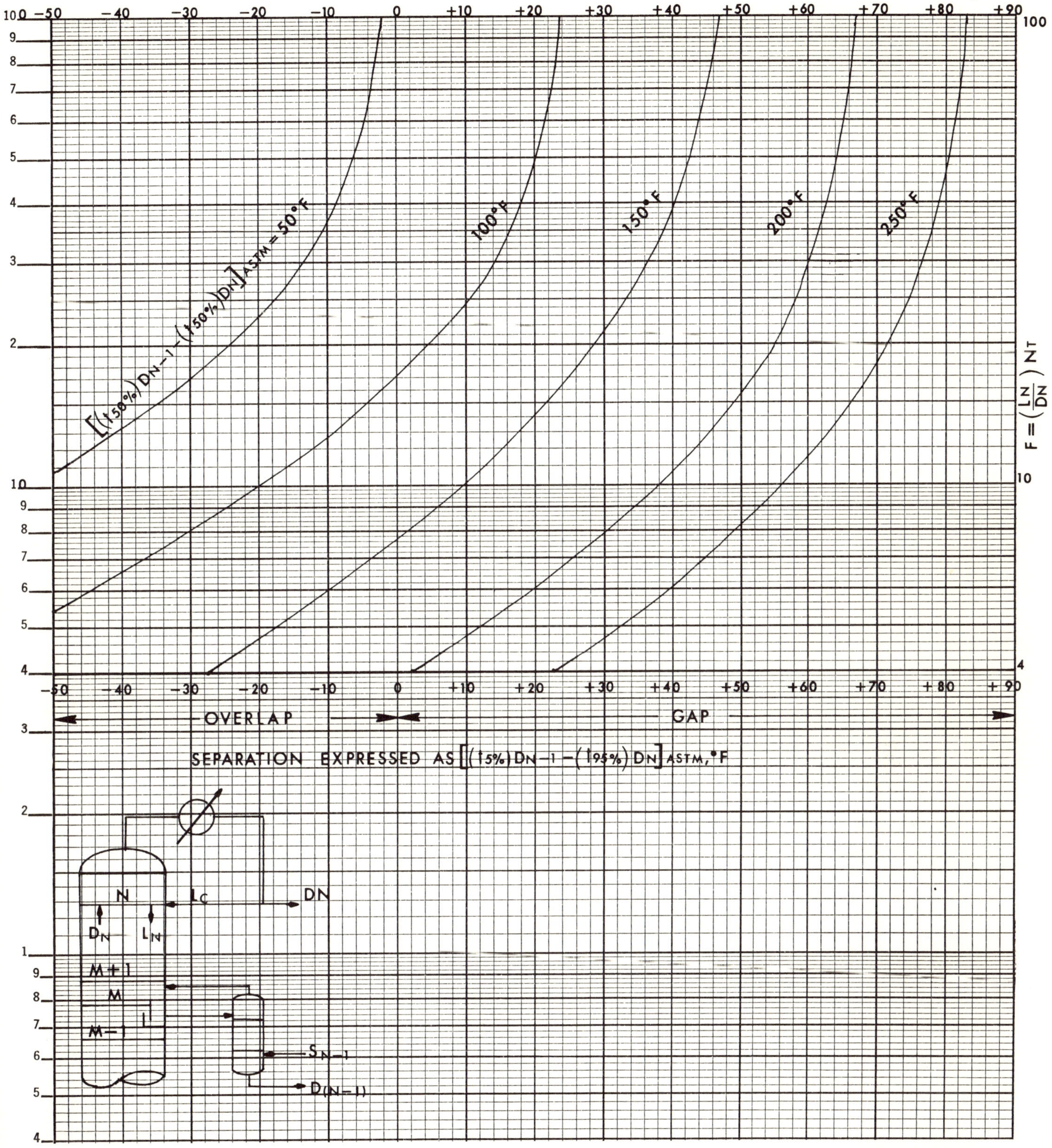

Figure 2.6. Fractionation between total overhead and highest sidestream product, atmospheric crude towers (used with permission of the American Institute of Chemical Engineers).

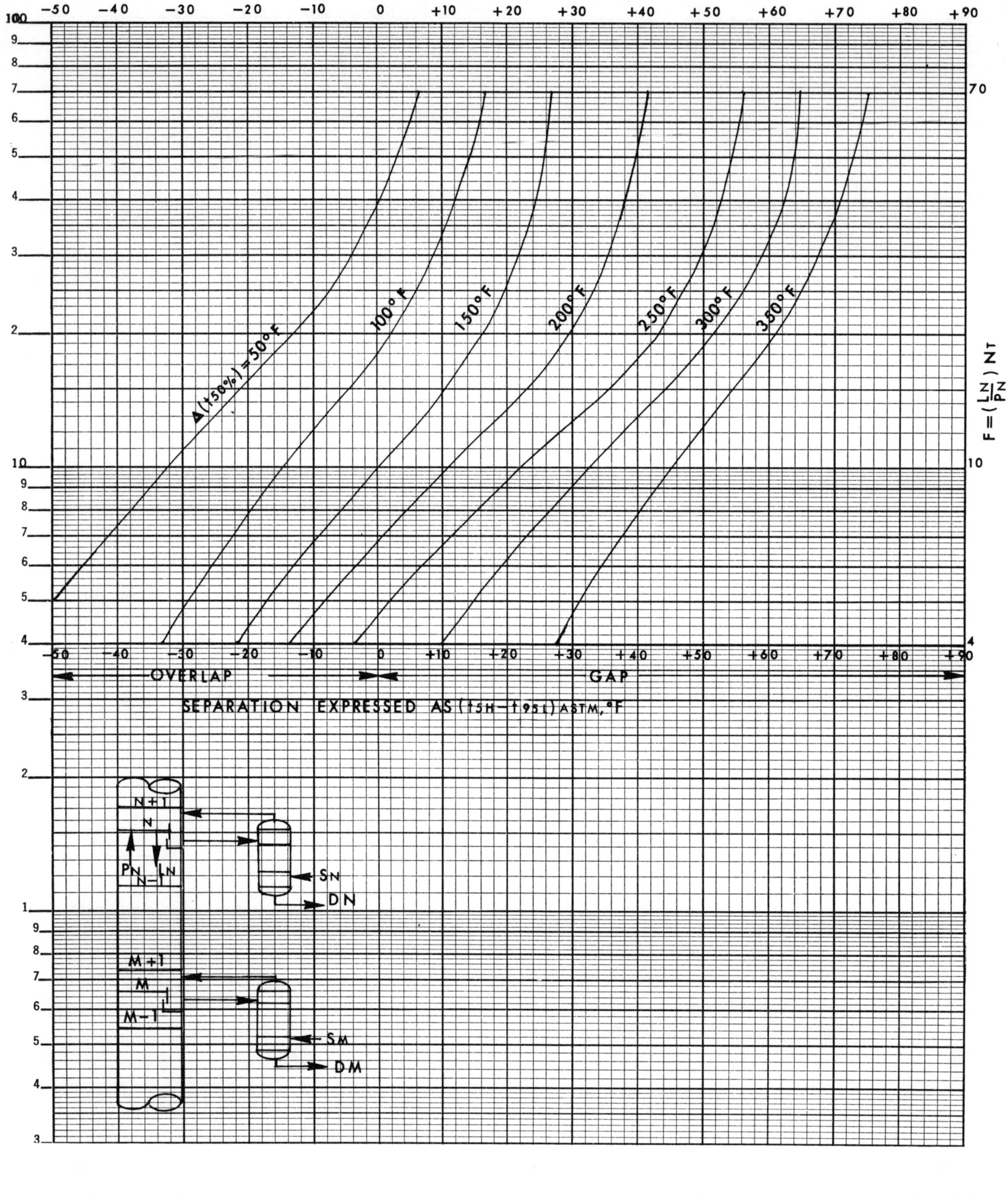

Figure 2.7. Fractionation between adjacent sidestream products, atmospheric crude towers (used with permission of the American Institute of Chemical Engineers).

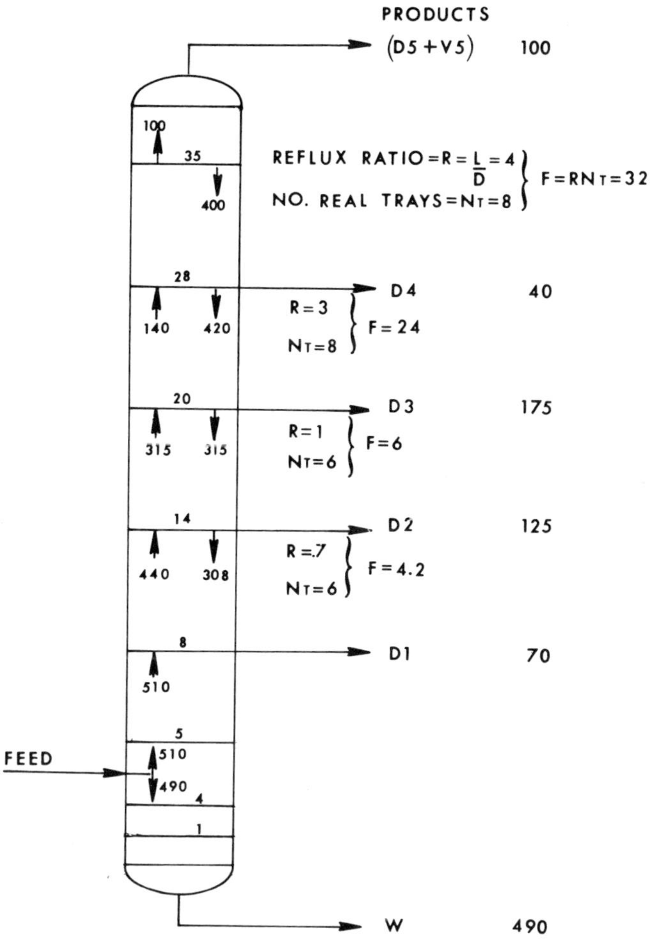

Figure 2.8. Product and reflux volumetric flow rates.

criterion is not met, Packie's curves for the case of insufficient steam stripping may be found in Nelson, pp. 515-516.

Packie defined the *degree of difficulty of separation* as the difference between the ASTM 50 volume percent temperatures of the fractions under study. For Figure 2.6, this difference is between the ASTM 50 percent temperatures of the first (highest) sidestream product and the overhead fraction. For Figure 2.7, this difference is between the ASTM 50 percent temperatures of the lower sidestream and the total remaining lighter distillates.

He then defined the *separation capability of the system* as the product of the reflux-to-feed ratio at the upper draw tray as calculated on a volumetric basis and the number of actual trays in the section. This product is designated as the *F-factor*. In sections where pumparound heat removal systems are used, trays in this service are considered to be only one-third of an actual fractionating tray.

Thus, Packie defines degree of separation attainable and the (5-95) Gap as functions of the separation capability of the system (F-Factor) with parameters of degree of difficulty of separation, Δt_{50} ASTM. To illustrate, consider three examples.

These examples are based on Figure 2.8 which illustrates the internal and external product rates and the internal reflux rates as expressed on a common volume/time basis and on Figure 2.9 which shows the volumetric product yield overlaid on the crude TBP curve. Figure 2.10, derived from Edmister's correlations (3), is used to convert the 50 volume percent TBP temperatures to ASTM tempertures.

The following conditions apply to Figure 2.8.

1. Vapor rates are products only measured as liquids at 60 degrees F.
2. Internal liquids and products are measured as liquids at 60 degrees F.
3. All flow rates are on a common volume/time basis.
4. Sidestream product strippers are not shown.
5. Separation

Separation	Tray Numbers in Section (Inclusive)	Number of Trays
D4 – (D5 + V5)	28 – 35	8
D3 – D4	20 – 27	8
D2 – D3	14 – 19	6
D1 – D2	8 – 13	6
W – D1	5 – 7	3

Example 1

D1 – D2 Separation—use Figure 2.7.
For D1,

(t_{50}) TBP = 703 degrees F – (Point A of Figure 2.9).

For D2 – D5,

(t_{50}) TBP = 410 degrees F – (Point B)

Converting these 50 percent temperatures from TBP to ASTM by using Figure 2.10,

Δt_{50} ASTM = 684 – 410 = 274 degrees F

F = 4.2

Therefore, from Figure 2.7, (5 – 95) Gap = + 6 degrees F.

Example 2

D3 – D4 Separation—use Figure 2.7.
For D3,

(t_{50}) TBP = 418 degrees F – (Point C)

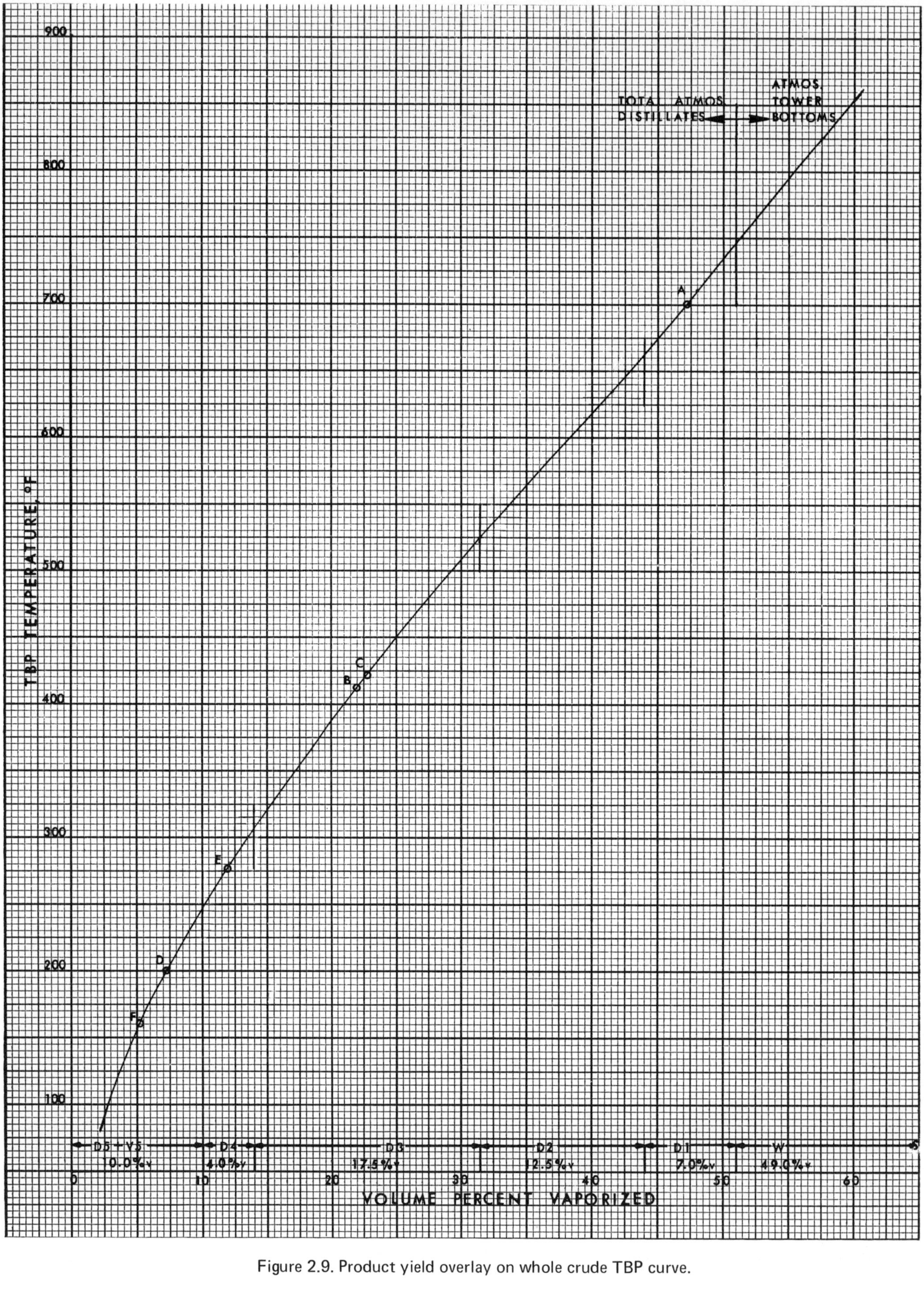

Figure 2.9. Product yield overlay on whole crude TBP curve.

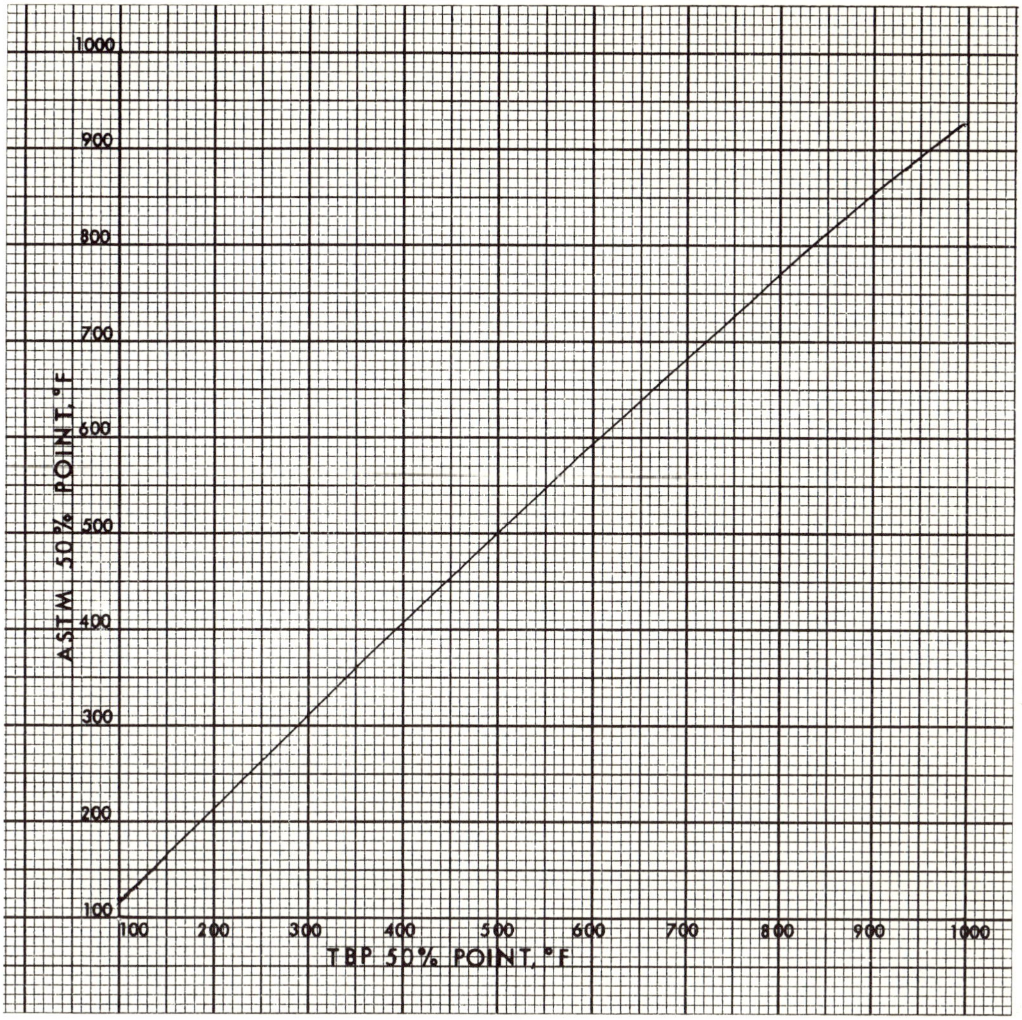

Figure 2.10. Relationship between ASTM and TBP 50 volume percent points.

For D4 – D5,

(t_{50}) TBP = 197 degrees F – (Point D)

Converting TBP to ASTM,

Δ t_{50} ASTM = 417 – 202 = 215 degrees F

F = 24

Therefore, from Figure 2.7, (5-95) Gap = +37 degrees F.

Example 3

D4 – D5 Separation–use Figure 2.6.
For D4,

(t_{50})TBP = 277 degrees F – (Point E)

For D5,

(t_{50})TBP = 159 degrees F – (Point F)

Converting TBP to ASTM,

Δt_{50} ASTM = 282 – 167 = 115 degrees F

F = 32

Therefore, from Figure 2.6, (5-95) Gap = +22 degrees F.

These computations will be encountered later in the example calculation but are introduced at this point to illustrate the principles of petroleum fractionation and its nomenclature. By inspection of Figures 2.6 and 2.7, it is clear that lowering the F-Factor, either by reducing the number of trays or the reflux ratio by altering the heat balance, will reduce the (5-95) Gap which is the separation. At this point, the similarity between discrete-component distillation and crude petroleum fractionation should be less unclear.

As more operating data has become available, Packie's work is now generally considered to be on the conservative side; that is, his procedure usually predicts a smaller gap than is actually attained in practice. The author believes that the degree of conservatism inherent in Packie is not excessive and recommends its use for design work. It is almost always true that crude distillation units are required to perform different operations than that for which they were designed. For this reason, the design "fat" contained in Packie becomes quite useful to the owner in later years.

Atmospheric Tower Charge Data

The feed to the atmospheric tower is the crude oil to be processed into the various products required. Before the design material balances can be developed, it will first be necessary to derive some basic physical property data for the crude. From this, one can then estimate the total distillate production and product distribution in the atmospheric tower.

In definitive process design work and/or in refinery operations analysis, the engineer usually has access to a complete crude assay. Indeed, he should not undertake a definitive design without one. On the other hand, feasibility studies and/or order-of-magnitude work is often performed with little more than a whole crude gravity and TBP curve. In either case, the first step is to characterize the crude in order to facilitate later calculations. The minimum information required is (a) whole-crude TBP curve, (b) whole-crude API gravity and (c) whole-crude light ends analysis.

Additional information which is highly desirable is an API gravity and molecular weight study of narrow cuts of the whole crude. From this can be derived a plot of volume percent versus weight percent and mid-volume percent versus molecular weight. If this data is not available experimentally, it can be calculated.

An experimental determination of the EFV curve, at least at atmospheric pressure and preferably at one or two higher pressures covering the anticipated range of operations is also desirable. Maxwell (4), Nelson and Edmister have all published procedures for converting a whole crude TBP curve to an atmospheric EFV curve and for estimating the EFV temperatures at pressures above atmospheric. In the petroleum industry, there is a good deal of argument about how to estimate EFV behavior of crudes since the total vaporization in the tower is the sum of the equilibrium flash vapor plus some stripout from the equilibrium flash liquid. Stripout versus stripping steam correlations are largely empirical. In an operating tower, one can calculate stripout by making a heat balance around the bottoms stripping section, but this is not possible in a design situation. For these reasons, the author strongly recommends obtaining experimental data, particularly when working with unfamiliar crudes. The cost of obtaining accurate equilibrium data is trivial when compared to the possible penalties involved in incorrectly estimating yields or missizing equipment.

However, if one is limited to the minimum data, the required information can be derived by using the following procedures for crude oil characterization.

Calculation Procedure for Characterizing Crude Oils

Given:

Whole-crude atmospheric TBP curve
Whole-crude gravity
Front end analysis for light ends

Calculate:

Volume percent versus weight percent
Mid-volume precent versus molecular weight

1. Calculate the characterization factor, K, for the whole crude using the techniques of Maxwell, Section 2.
   a. Calculate the volume average boiling point (VABP) using the 20, 50 and 80 volume percent TBP temperatures.
   b. Calculate the 10 to 70 slope of the whole-crude TBP curve.
   c. Using the proper correction factor, convert VABP to mean ABP.
   d. K is found as a function of mean ABP and API gravity by use of Winns' nomogram (5).
2. Assuming that K remains constant, calculate the molecular weight and API gravity for various boiling-range cuts.
   a. For the TBP range (initial boiling point to 200 degrees F), calculate mean ABP, API gravity and molecular weight. Based on 100 barrels of whole crude, calculate and tabulate:

      barrels vapor
      pounds vapor
      moles vapor
      molecular weight of vapor

   b. For the TBP range (200 to 300 degrees F), repeat Step 2a.
   c. For the TBP range (300 to 400 degrees F), repeat Step 2a, and so on until the entire crude range has been covered.

d. From these calculations, draw the following curves.
   1. Volume percent over versus weight percent over.
   2. Mid-volume percent versus molecular weight of vapor.
3. Calculate the atmospheric pressure EFV curve for the whole crude using Packie's procedure. Plot this curve on the same chart as the whole-crude TBP curve.

Estimate of Material Balance

As is the case in any process-design problem, the definition of the material balance is the first and most important step. In this portion of the discussion, various methods for estimating product yields from the crude petroleum feed will be explored.

The design material balance is determined by the product characteristics required by the owner and by the amount of crude vaporization which will occur at the conditions of temperature and pressure existing in the flash zone.

Since most crude distillation units contain both atmospheric and vacuum towers, economic considerations usually favor maximizing distillate yield from the atmospheric section in order to minimize the load on the vacuum section. This may not always be true in an operating situation where the vacuum tower might have unused capacity when the atmospheric tower is operated for maximum distillate production. In that case, one would balance atmospheric and vacuum distillate production in such a way as to maximize crude throughput. In designing atmospheric towers where the bottoms liquid is sold directly as fuel oil, the most economical approach in this special case is to vaporize only the required distillate products and to allow the remaining potential distillate to be yielded, unvaporized, with the bottoms. This situation arises often in foreign refineries, usually in the producing areas, which run crude primarily for local fuels requirements and for bunkering tankers. Thus, there is just no economic driving force for yielding anything heavier than diesel fuel as a distillate.

In practice, the question of optimizing the relative distillate yield between the atmospheric and vacuum towers will be settled on an economic basis and must be resolved prior to commencing definitive design work. This type of analysis will depend upon economic factors within the particular company and/or plant site under study and is outside the scope of this work.

Total Distillate Yield

The total distillate yield is found by calculating the vaporization of the crude which will occur at the conditions of temperature and hydrocarbon partial pressure existing in the flash zone. The principles of this analysis are discussed as follows.

TABLE 2.1

Tray Requirements in Atmospheric Crude Towers

Separation	Number of Trays
Light naphtha to heavy naphtha	6 to 8
Heavy naphtha to light distillate	6 to 8
Light distillate to heavy distillate	4 to 6
Heavy distillate to atmospheric gas oil	4 to 6
Flash zone to first draw tray	3 to 4
Steam and reboiled stripping sections	4

Flash Zone Pressure

Flash zone pressure should be set at the minimum possible level. This will maximize the crude vaporization or, if less than maximum vaporization is desired, will minimize the operating temperature level.

The datum pressure in atmospheric tower design is the lowest permissible pressure in the towers' reflux accumulator. Since, in many cases, the tower overhead product stream cannot be totally condensed at a reasonably low pressure, there will often be a vapor distillate which must be removed from the tower system and sent either to the plant fuel gas system or to further processing. This is normally done by a compressor located as near as possible to the tower. In most refineries, this distillate vapor is fed to a gas plant for recovery of propane and heavier. In refineries running heavy crudes, the volume of recoverable light ends may be too small to justify recovery facilities. In this case, the distillate vapor is sent to the plant fuel gas system. The author has observed one operation where the distillate vapor was flared. This required a considerably higher reflux drum pressure than if compressors were used to feed other systems. Admittedly, this was a rare operation. For a grassroots design a reflux drum pressure as low as 0.2 psig is practical, provided that suction lines to the compressor can be kept short. If the installation involves tying a new tower into an existing compressor which may be located a considerable distance away, a higher pressure may be required. The normal range of reflux drum pressures used is 0.2 to 1.0 psig.

The design pressure drop across the condenser(s) and associated piping should be set at 5 to 7 psi. When the system is clean, the normal pressure drop will be on the

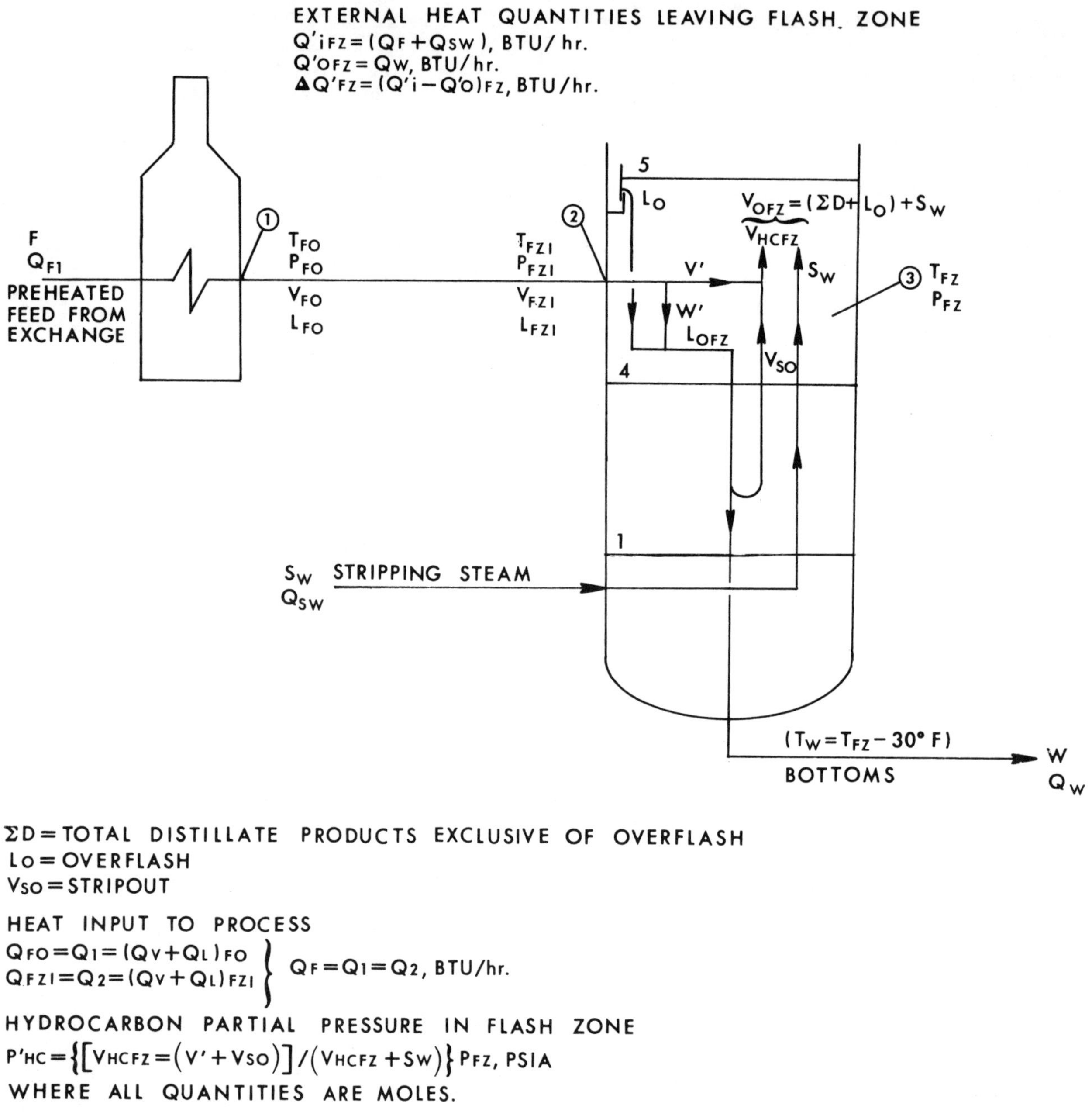

Figure 2.11. Flash zone heat and material balance.

order of 3 psi. However, as the exchangers become fouled, the extra drop allowed will be available.

Most atmospheric towers have 25 to 35 trays between the flash zone and the tower top. Recommended ranges for the number of trays in various sections of the tower are given in Table 2.1. Using these numbers of trays and reasonable levels of heat recovery will provide adequate fractionation between products. The allowable pressure drop for trays will be in the range of 0.1 to 0.2 psi per tray. In the design example, a pressure drop of 5 psi was taken between the flash zone and the tower top.

At this point, the flash zone pressure is set as the sum of the reflux drum pressure and the combined pressure drops across the condensers and trays above the flash zone. A pressure drop of 5 psi between the flash zone and the furnace outlet is recommended.

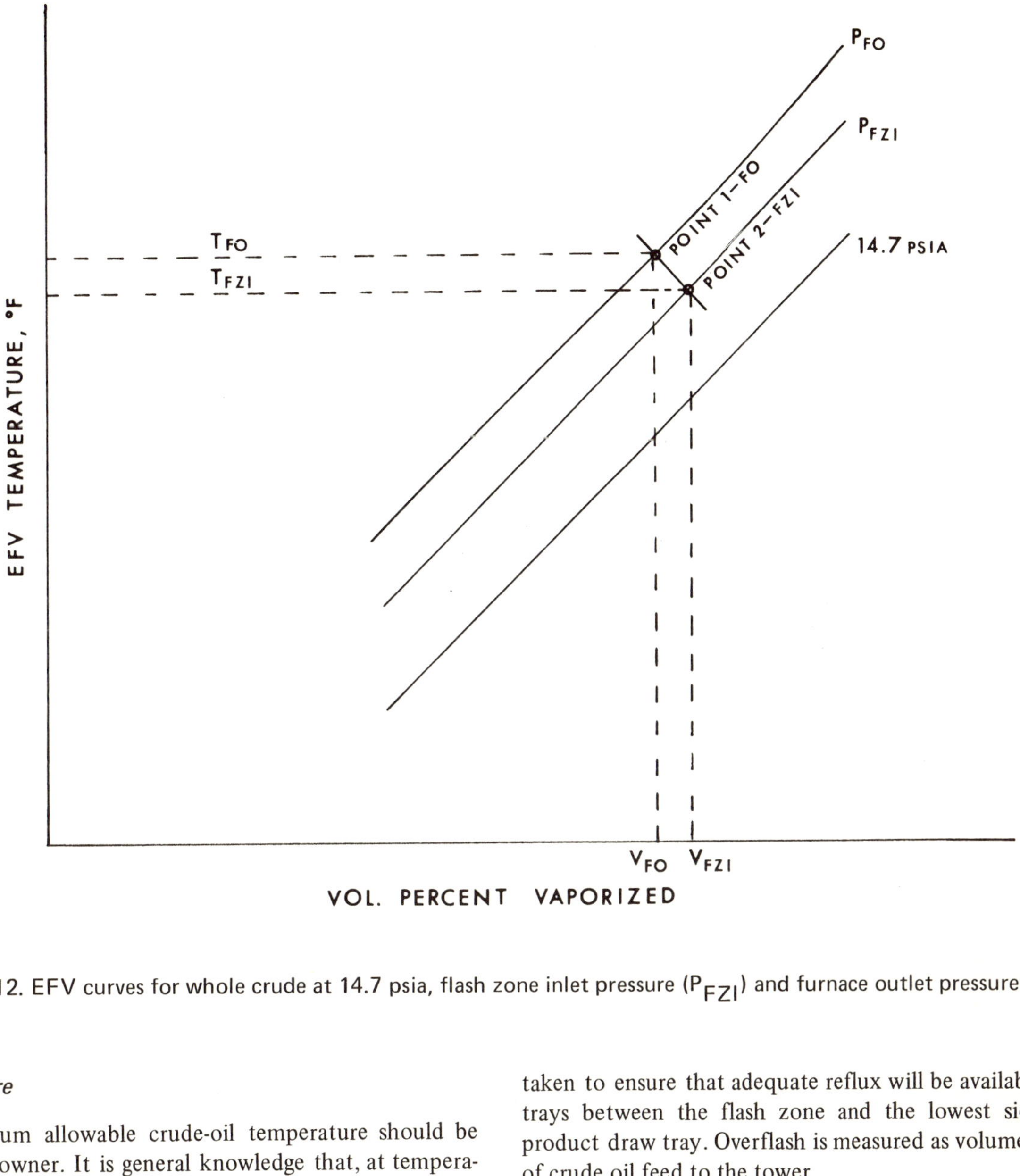

Figure 2.12. EFV curves for whole crude at 14.7 psia, flash zone inlet pressure (P_{FZI}) and furnace outlet pressure (P_{FO}).

Temperature

Maximum allowable crude-oil temperature should be set by the owner. It is general knowledge that, at temperatures of 650 to 675 degrees F, most paraffinic and naphthenic base crudes begin to undergo thermal decomposition. This contributes to poor color and poor smoke properties in the light and heavy distillate fractions. Although oil temperatures as high as 735 degrees F for kerosene operation and 800 degrees F for general operations have been reported (6), the author recommends that temperatures of the oil should not exceed 650 degrees F for kerosene production and 700 degrees F for other designs.

Overflash

Overflash is an extra amount of vaporization above that required by material balance considerations which is taken to ensure that adequate reflux will be available in the trays between the flash zone and the lowest sidestream product draw tray. Overflash is measured as volume percent of crude oil feed to the tower.

The value of overflash to be used in design has been the subject of much discussion through the petroleum industry. An overflash of 2 volume percent of feed is recommended for competitive design although some refiners will specify values as high as 5 volume percent. On the other end of the scale, the author has observed a successful although very sensitive tower operation at overflash values as low as 0.2 volume percent. Since the gross hydrocarbon vapor from the flash zone is potentially all product, an actual operation will allow as much of the specified overflash as possible to come up into the lowest sidestream, rejecting only that amount necessary to control color in the distillates by refluxing the trays above the flash zone.

A conservative approach to design and one providing maximum operating flexibility would be to design the atmospheric section on the basis of maximum oil temperature and zero overflash. This means that the lowest sidestream product could contain all the overflash. Accordingly, this would require that the heat exchangers in the lowest sidestream circuit be sized to accommodate this high potential rate. On the other hand, the vacuum section would be designed on the basis that the atmospheric tower would return the design value of overflash to the reduced crude. This would require additional vacuum furnace capacity and sizing the heat exchangers in the lightest vacuum gas oil circuit for the highest potential rate. Because of the significantly higher cost, this approach to crude unit design would seem indicated only where a high degree of operating flexibility is required and where the penalties for failure to meet product quality standards are severe. For competitive design work, this approach would lead to financial disaster.

Calculation of the Total Distillate Yield

The calculations outlined in this subsection are based on having predetermined the maximum temperature of the oil leaving the furnace. Occasionally, one may define the thermal basis as the flash zone temperature. If this is the basis, the procedure described in the chapter covering the vacuum tower–specifically Figure 3.8–will apply for flash zone calculations.

Figure 2.11 illustrates the flow rates around the flash zone and gives the equations used in defining the appropriate heat quantities. Figure 2.12 shows how the basic heat input to the process is calculated by using the EFV curves. The sequential calculation procedure is as follows.

1. Convert the 14.7 psia EFV curves to the pressures which have been set for the furnace outlet and the entrance to the flash zone. This latter value is, for all practical purposes, equal to the flash zone pressure. The vapor pressure charts and nomographs in Section 4 of Maxwell are useful for these conversions.
2. At the furnace outlet conditions of temperature and pressure, find the volume percent vaporized as Point 1 on Figure 2.12. From crude assay data, calculate the weight of the vaporized crude. Using the weight of the vapor and liquid phases, calculate the heat content of the mixture leaving the furnace as

   $$Q_1 = Q_{FO}$$

3. Next, determine the vaporization which exists at the flash zone entrance as Point 2 on Figure 2.12. This is a trial-and-error procedure which requires assuming a new temperature at the flash zone entrance, determining the new percentage vaporization and then calculating the heat content until satisfying the criterion

   $$Q_1 = Q_2 \text{ or } Q_{FO} = Q_{FZI}$$

   The temperature drop between the furnace outlet and the flash zone inlet will usually be on the order of 4 to 6 degrees F. The amount of heat which the feed introduces into the process has now been defined as

   $$Q_F = Q_{FO} = Q_{FZI}$$

4. Set the value for overflash. This liquid returns to the flash zone from the tray immediately above it and mixes with the feed flash liquid. The sum of these two liquids is the feed to the top tray of the bottoms stripping section.
5. Set a stripping steam rate–10 pounds per barrel of net bottoms is recommended. From Figure 2.13, find the volume percentage of the feed to the stripping section which will be vaporized. From this difference, calculate the volume of net bottoms. From crude assay data, calculate the weight of the bottoms product.
6. The total vapor leaving the flash zone is the sum of the feed flash vapor and the stripout. From this is calculated the volume, weight and molal quantities around the flash zone.
7. Calculate the hydrocarbon partial pressure in the total vapor leaving the flash zone as P'_{HC}. Find the 14.7 psia EFV temperature for the indicated volume percentage vaporization and convert this to the pressure level of P'_{HC}. *This is the flash-zone temperature.*
8. Assume that the temperature of the net bottoms liquid, W, is 30 degrees F lower than the flash zone temperature. Calculate the heat quantity leaving the system in this stream.
9. Calculate the heat input to the system from the stripping steam.
10. Calculate and tabulate the external heat quantities at the flash zone.
11. Tabulate the quantities and properties of the hydrocarbon flows around the flash zone.
12. Total distillate yield is the total hydrocarbon vapor leaving the flash zone minus the overflash.

A typical calculation of total distillate yield is given in the appended example calculations.

General Properties of Petroleum Fractions

Most petroleum distillates, especially those from atmospheric distillation, are usually defined in terms of an

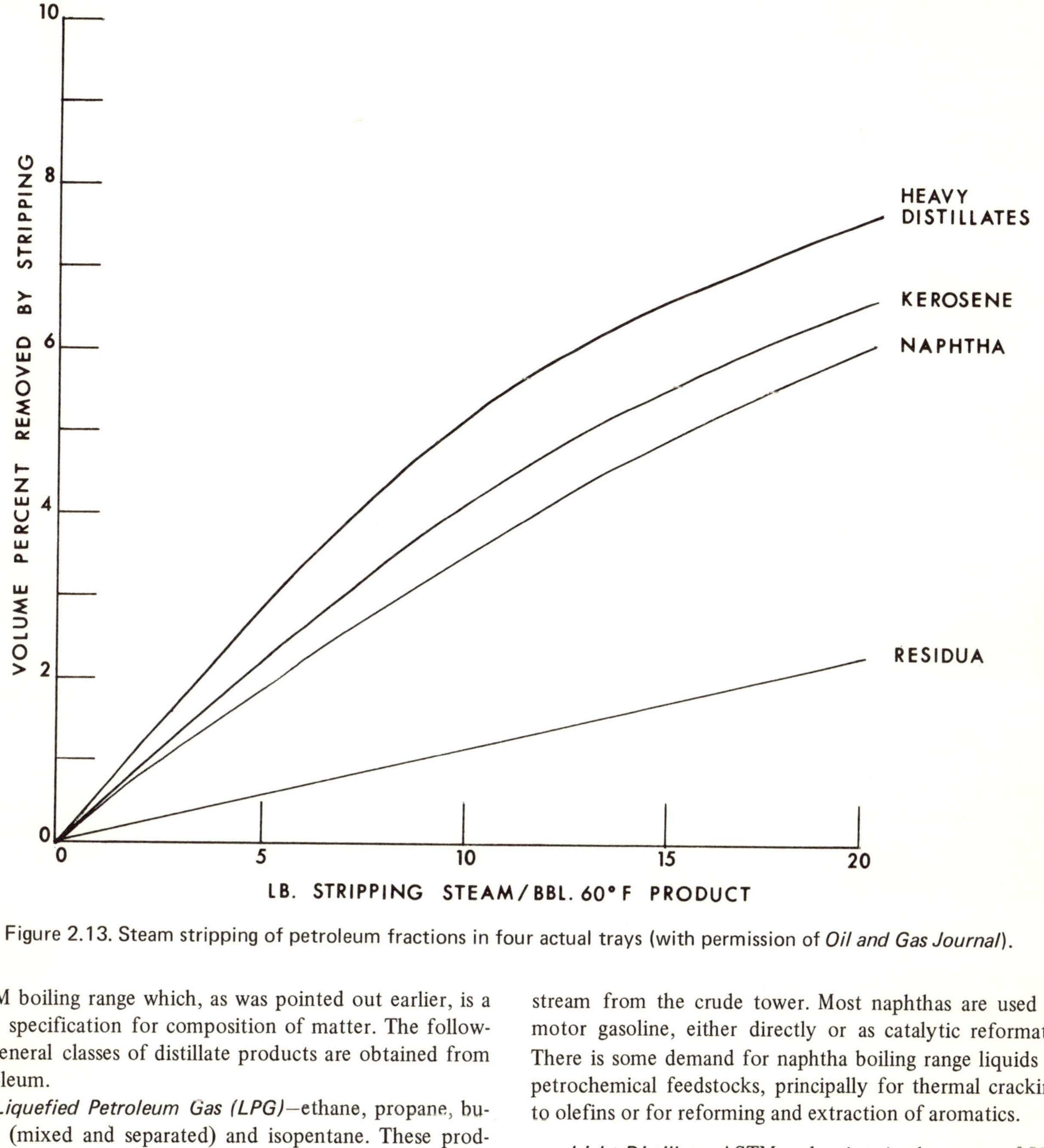

Figure 2.13. Steam stripping of petroleum fractions in four actual trays (with permission of *Oil and Gas Journal*).

ASTM boiling range which, as was pointed out earlier, is a loose specification for composition of matter. The following general classes of distillate products are obtained from petroleum.

Liquefied Petroleum Gas (LPG)—ethane, propane, butanes (mixed and separated) and isopentane. These products cannot be produced directly from crude distillation and require further high pressure distillation.

Naphtha (Gasoline)—pentane through 400 degrees F ASTM end point. There are many grades and boiling ranges of naphthas, the properties of which are influenced by their planned end use. Many refineries produce a 400 degree F end point naphtha as an overhead distillate from the atmospheric tower and then fractionate it as required in separate facilities. If the product pattern can be defined well enough it is more economical to produce a light naphtha as the overhead distillate and a heavy naphtha as the lightest sidestream from the crude tower. Most naphthas are used in motor gasoline, either directly or as catalytic reformate. There is some demand for naphtha boiling range liquids as petrochemical feedstocks, principally for thermal cracking to olefins or for reforming and extraction of aromatics.

Light Distillate—ASTM end points in the range of 525 to 575 degrees F. A full range light distillate is kerosene having an ASTM boiling range of 325 to 550 degrees F. An important distillate is JP-4 jet fuel which has an ASTM boiling range of approximately 150 to 475 degrees F. This product is blended from suitable stocks of naphtha and light distillate. However, under proper operating conditions, this material could be yielded directly from the atmospheric tower as a light distillate.

Heavy Distillate—ASTM end points in the range of 650 to 700 degrees F. Diesel fuel is a blend of light distillate and

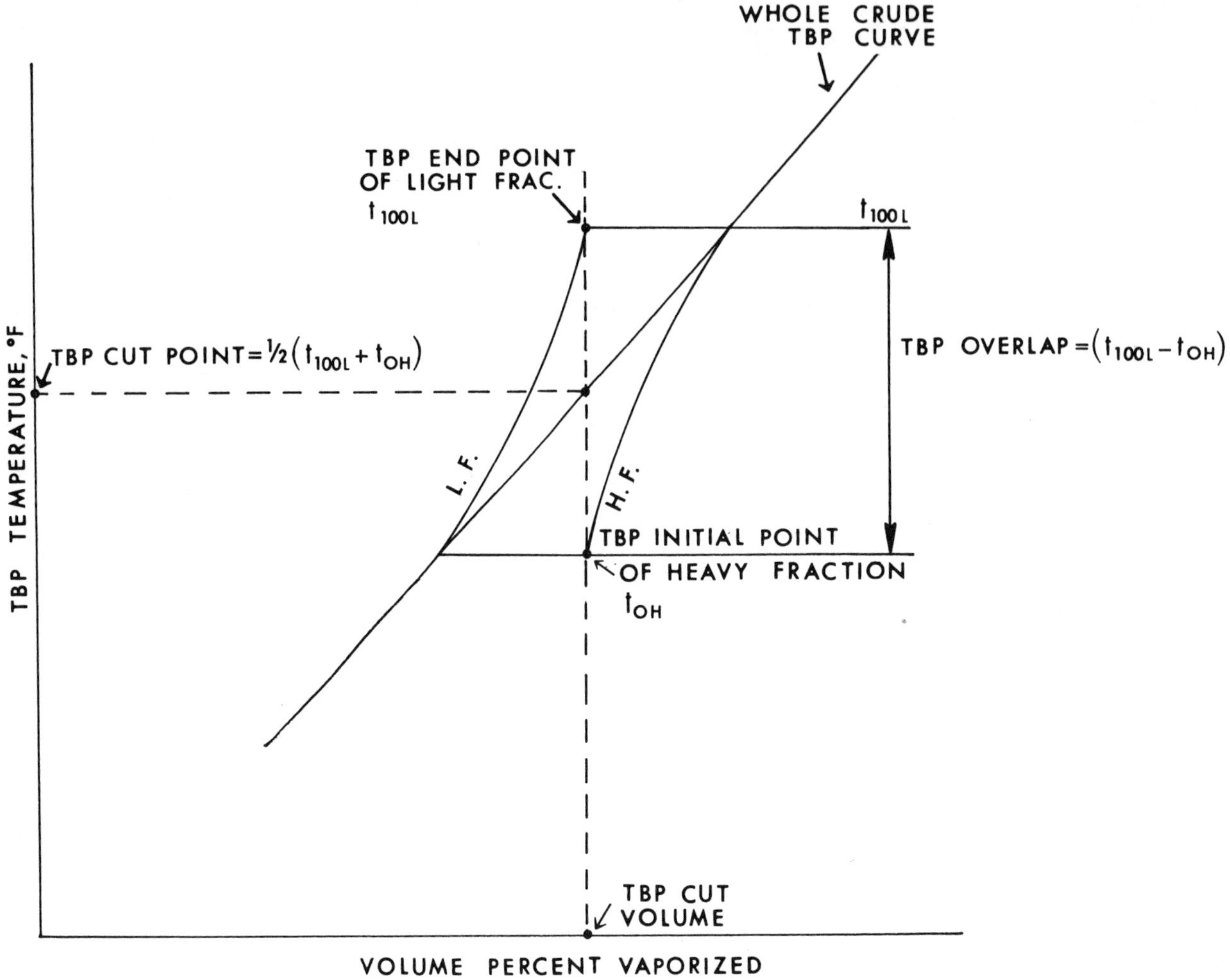

Figure 2.14. Temperature relationships around the cut point between fractions.

heavy distillate and has an ASTM boiling range of approximately (350 to 675 degrees F). Marine diesel is a little heavier, having an ASTM end point in the range of 775 degrees F.

Gas Oil—all distillates heavier than heavy distillate. Gas oil yielded from the atmospheric tower will have ASTM end points of approximately 800 degrees F. Vacuum gas oils will have ASTM end points as high as 1,100 degrees F. Depending upon the operation being practiced, vacuum gas oils can have various boiling ranges. This subject is covered in detail in Chapter 3.

Note that the boiling ranges of these materials often overlap by quite a bit. This indicates that one can maximize the yield of only one product on any given operation. For example, consider maximizing the production of 325 to 525 degrees F kerosene. It is obvious that the naphtha end point cannot be 400 degrees F—corresponding to a maximum naphtha operation—nor can the heavy distillate initial point be much lower than 500 degrees F which precludes a maximum heavy distillate operation. The importance of this principle will be seen in the following material where methods for cutting the crude into products are presented.

Product Distillate Yields

In this part will be discussed methods for estimating the yields and properties of the various products which can be produced off the atmospheric tower. The general method of attack will be to determine distillation properties of the desired products and apply these to the whole crude TBP curve in order to estimate volumetric yields. Volumetric yields are then used to obtain product gravities, molecular weights and other properties from the crude assay. There are many ways to define a product slate, but most of these are based on specifying certain ASTM distillation tem-

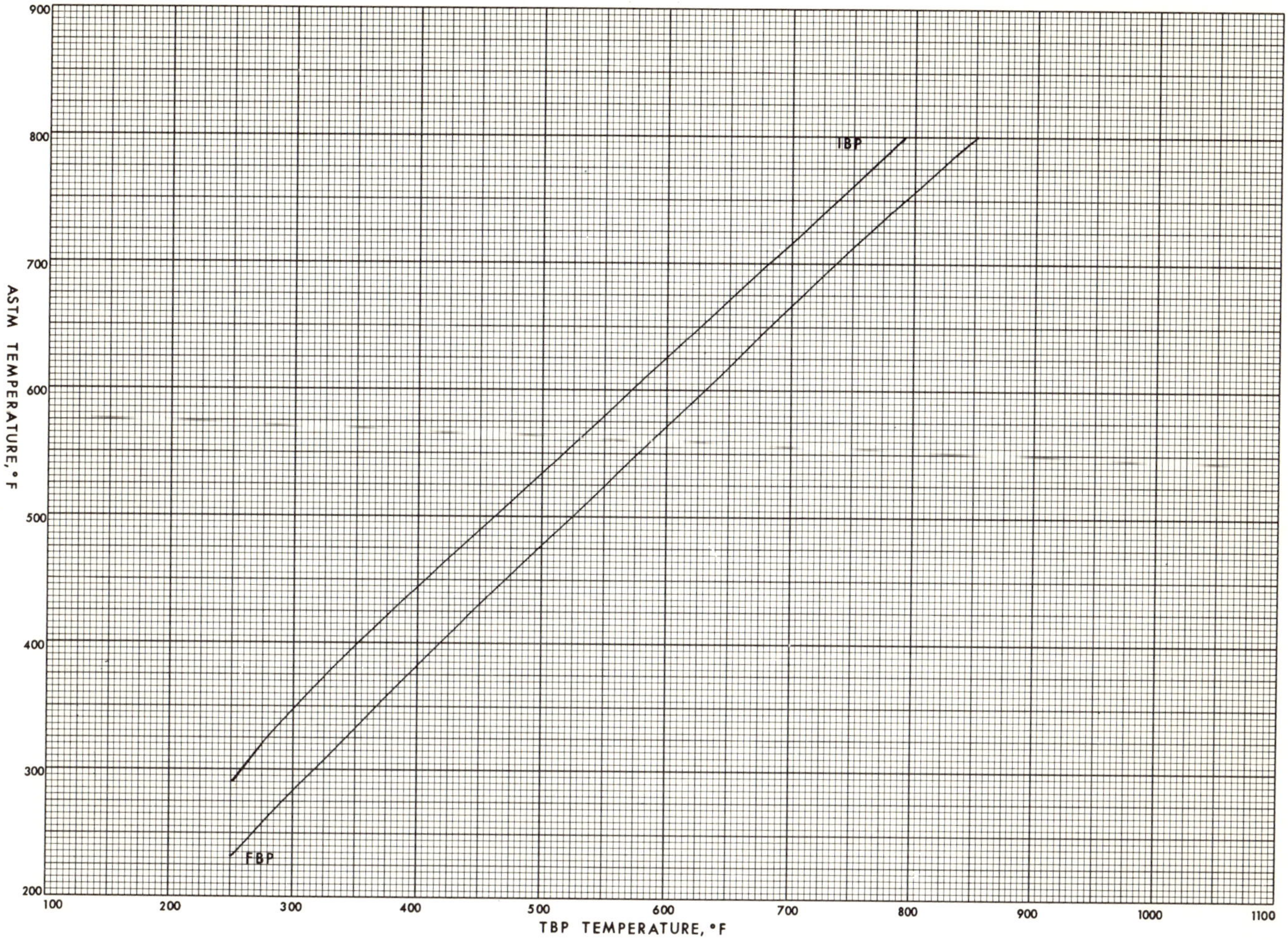

Figure 2.15. Relationships between ASTM and TBP initial (0 percent) and final (100 percent) boiling points.

peratures for the products from which the remaining data will be derived.

Since the planned production rates will depend upon the crudes which are available to the refinery, the designer should base his yield studies on the heaviest (lowest API gravity) and the lightest (highest API gravity) crude which is to be run on a normal basis. Further, he should obtain specific definition of all the operations which are to be practiced. These two points are very important and are discussed in detail in the following paragraphs.

The light crude will define the design basis for the atmospheric section of the crude unit since its volume of distillates will exceed that which can be produced from the heavy crude. All equipment sizing will be based on heat and material balance data calculated for the various light crude cases. As would be expected, the heavy crude will define the facilities for processing the atmospheric tower bottoms, either a vacuum unit or, if this is not planned, the reduced crude heat exchange equipment. To further complicate the matter, the various production operations must be calculated—heat and material balance—for both crudes. This will usually entail the alternate maximization of naphtha, light distillate and heavy distillate. These studies will define the variations in the heat and material balance required to satisfy the varying yields pattern and are necessary for the optimum design of the heat exchange train, towers and furnaces.

Now, the ways in which one can define yields and, hence, the overall material balance are to be considered. It should be remembered that, to this point, the hydrocarbon vapor leaving the flash zone and the overflash have been calculated. Thus, the total distillate yield is known. In discussing the various ways for estimating yields, the following terminology will be used. Figure 2-14 illustrates the physical significance of these terms.

1. *TBP cut volume*—the volumetric yield point between two fractions.

TABLE 2.2

Recommended ASTM Boiling Ranges for Products Atmospheric Tower			
Product	Max. Naphtha	Operation Max Light Dist.	Max Heavy Dist.
Light naphtha	250 – 275 EP	Same	Same
Heavy naphtha	400 EP	325 EP	325 EP
Light distillate	375 – 600	300 – 600	300 – 550
Heavy distillate	575 – 675	575 – 675	525 – 675
Atmospheric gas oil	Determined by allowable oil temperature		

TABLE 2.3

Separation Criteria for Atmospheric Tower Products	
Separation	(5 – 95) Gap, °F
Light naphtha to heavy naphtha	+ 20 to + 30
Heavy naphtha to light distillate	+ 25 to + 50
Light distillate to heavy distillate	0 to + 10
Heavy distillate to atmospheric gas oil	0 to + 10

2. *TBP cut point*—the whole crude TBP temperature corresponding to the TBP cut volume.
3. *TBP overlap* = (TBP end point of light fraction)–(TBP initial point of heavy fraction).

Also, note that

$$\text{TBP cut point} = \tfrac{1}{2}(t_{100L} + t_{OH}).$$

This latter principle has been established by analysis of operating data which has shown that, for well stripped sidestreams, the volume interchanges between two streams around the TBP cut point are equal.

Crude Oil Inspection Data

The owner defines the material balance himself by specifying the volumetric yields on crude for all products down through heavy distillate. The designer will determine the yield of atmospheric gas oil. This type of specification will almost always include the anticipated ASTM distillations and other key physical properties of the streams. This data should be checked for accuracy and any discrepancies resolved before proceeding with the design.

ASTM Distillations of Products

The owner specifies the ASTM distillations for the sidestreams down through heavy distillate and the ASTM end point of the overhead product. He may also specify the ASTM initial point of the atmospheric gas oil, but, if not, this shall be determined by the designer. The designer must also determine the TBP cut point between the overhead and the lightest sidestream. These specifications are used to estimate yields by the following procedure.

1. For the sidestreams, convert the ASTM to TBP distillations by Edmister's techniques.
2. The ASTM end point of the overhead is converted to a TBP end point by the correlation of Figure 2.15 (derived by author).
3. If the ASTM initial point of the atmospheric gas oil has been given, convert it to a TBP initial point using Figure 2.15. If this ASTM temperature has not been given, assume a TBP overlap of 80 to 100 degrees F between heavy distillate and atmospheric gas oil.
4. Calculate the TBP cut points between fractions and determine the volumetric yields of products.

Key Stream Specification

The owner desires that the design be based on production of one particular stream and that the other products be defined in terms relative to but subordinate to the key stream. Usually, this specification will give detailed ASTM temperatures for the key stream. The 0, 5, 10, 50, 90, 95 and 100 volume percent ASTM temperatures are the ones most often used along with (5-95) Gap specifications between the key stream and adjacent streams.

General Specifications

In the unlikely event that work is to be undertaken without having specific instructions from the owner, the following procedure is recommended.

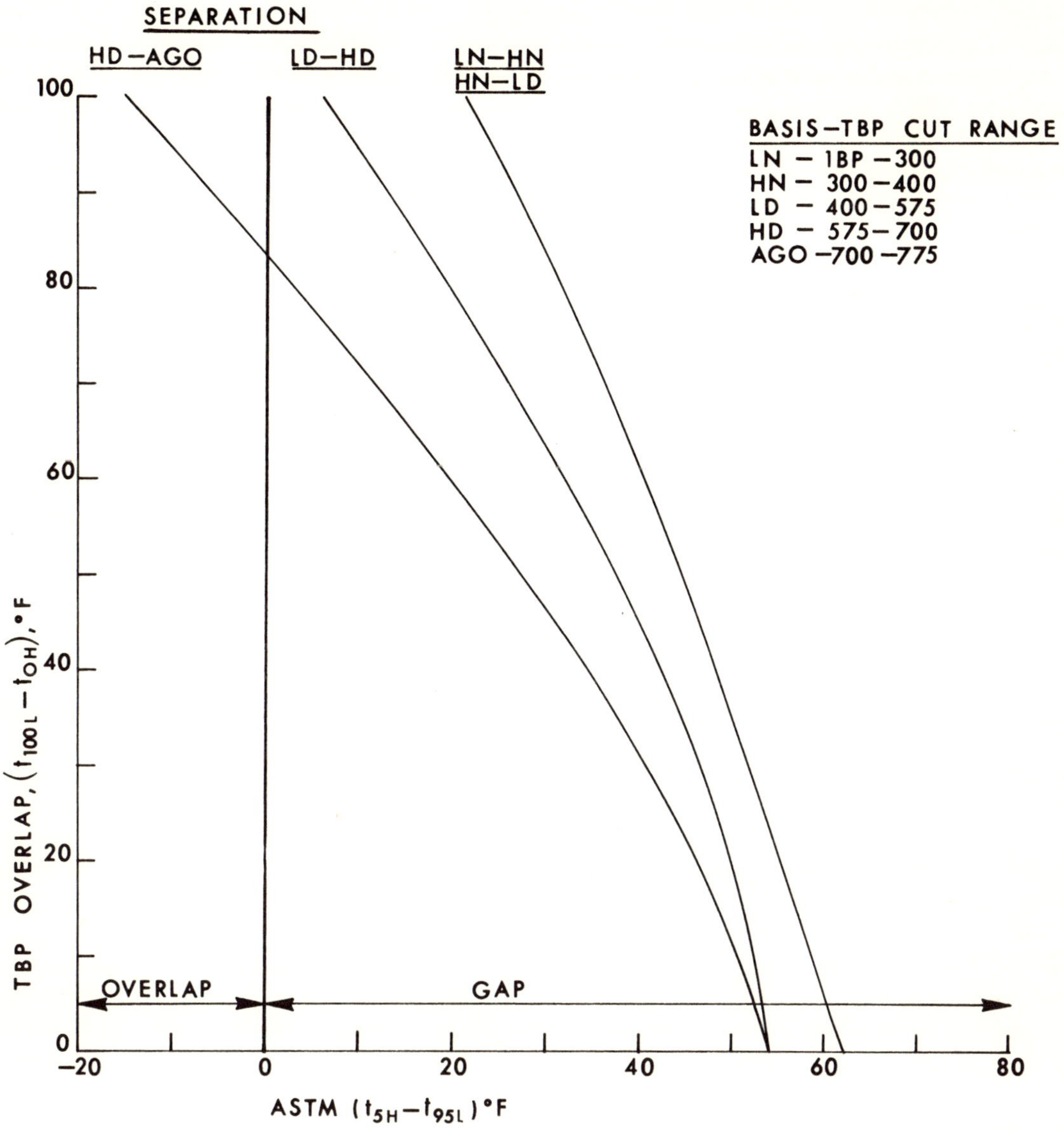

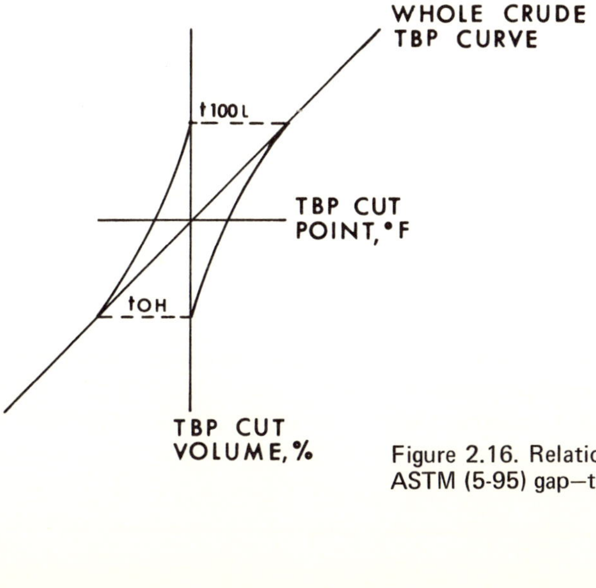

Figure 2.16. Relationships between TBP (100-) overlap and ASTM (5-95) gap—typical atmospheric tower streams.

1. As discussed earlier, the material balance should be based on two representative crudes, one light and one heavy. The material balances will then be based on alternately maximizing naphtha, light distillate and heavy distillate production. Total distillate yield is based on a maximum oil temperature leaving the furnace of 700 degrees F. Suggested ASTM boiling ranges for these cases are given in Table 2.2.
2. Recommended separation criteria, i.e., ASTM (5-95) Gap, are given in Table 2.3.

Figure 2.16 may be used to estimate TBP overlap for a given ASTM (5-95) Gap between the indicated fractions.

Product Properties

To this point, the volumetric yields of all products and the TBP and ASTM distillations for all distillates have been estimated. In order to complete the estimate of the material balance, it is necessary to define various other properties of the materials. The following steps will accomplish this.

1. Plot the ASTM curves of the distillates and calculate the (5-95) Gaps.
2. Calculate the 14.7 psia EFV curves for the distillates. Plot these curves and extrapolate to minus 20 volume percent vaporized in order to approximate the effect of the equilibrium solubility of the lighter crude oil components in the liquid leaving the draw tray.
3. Set stripping steam to the atmospheric gas oil stripper at 10 pounds per barrel stripped product and estimate the stripout from Figure 2.13. Since the other sidestreams are to be reboiled, set their vaporizations equivalent to the amount which would be stripped out by steam at a rate of 10 pounds per barrel stripped product. Plot these stripout points, i.e., minus vaporization, on the product EFV curves. These are the 14.7 psia bubble points of the unstripped sidestream products and will be used later to calculate draw tray temperatures.
4. From the crude assay data, calculate the gravities and molecular weights of all products.
5. Calculate the vapor-liquid separation of the gross overhead product at the conditions of temperature and hydrocarbon partial pressure existing in the reflux drum. These calculations are detailed in the design example. The detailed calculation procedure is as follows.
   a. The light ends analysis and the partial TBP curve of the gross overhead is combined into a total TBP curve.
   b. The TBP curve is broken up into pseudocomponents which are tabulated as volume, weight and molal quantities. The use of n-alkane physical properties is acceptable.
3. Since free water will exist in the reflux drum, the vapor phase will be water saturated at the condenser exit temperature, T_R. Calculate the hydrocarbon partial pressure of the distillate vapor stream as P'_{HC}.
   d. At conditions T_R and P'_{HC}, make a flash calculation on the gross overhead stream, thereby defining the vapor-liquid separation. From this, define the composition of the vapor and the liquid.
   e. Synthesize a TBP curve for the distillate liquid and convert it to an ASTM distillation. This step is omitted in the example calculation.
   f. Calculate and tabulate the distribution of the steam to the process as vapor leaving with the distillate vapor and liquid leaving as free water.
   g. Assume the number of trays for the various separation sections in the tower and define the total number of trays and draw tray locations. Plot the gravity and molecular weights of the liquids leaving trays by assuming that the properties of liquids one tray above and one tray below draw trays are the same as the draw tray liquid. Plant test work has indicated that this is true. Assume a linear change in properties across the remaining trays in each section. This plot is used only in the calculation of Type A systems.

Process Design Basis

The material balance and composition information which has been developed should now be tabulated into a process design basis for ready reference in later calculations.

Heat and Material Balance Calculations for Type U Towers

A complete Type U tower is shown in Figure 2.17. This drawing illustrates the basic process and its essential auxiliaries as well as the external heat and material balance quantities. Note that the product draw trays are all shown as partial draws so that the reflux flows internally from the draw tray to the tray below it. Figure 2.17 will be the basis for discussing the heat and material balance calculations in this section. In the introductory remarks to this chapter, it was stated that a Type U system is not practical from an industrial viewpoint because of the obvious lack of thermal efficiency. However, it is the easiest system to calculate and to understand. Thus, the fundamentals of heat and material balance calculations will be illustrated in terms of a Type U system. Later, the methods for obtaining and calculating heat removals from the tower will be given. A secondary reason for calculating a Type U system is that, as will be

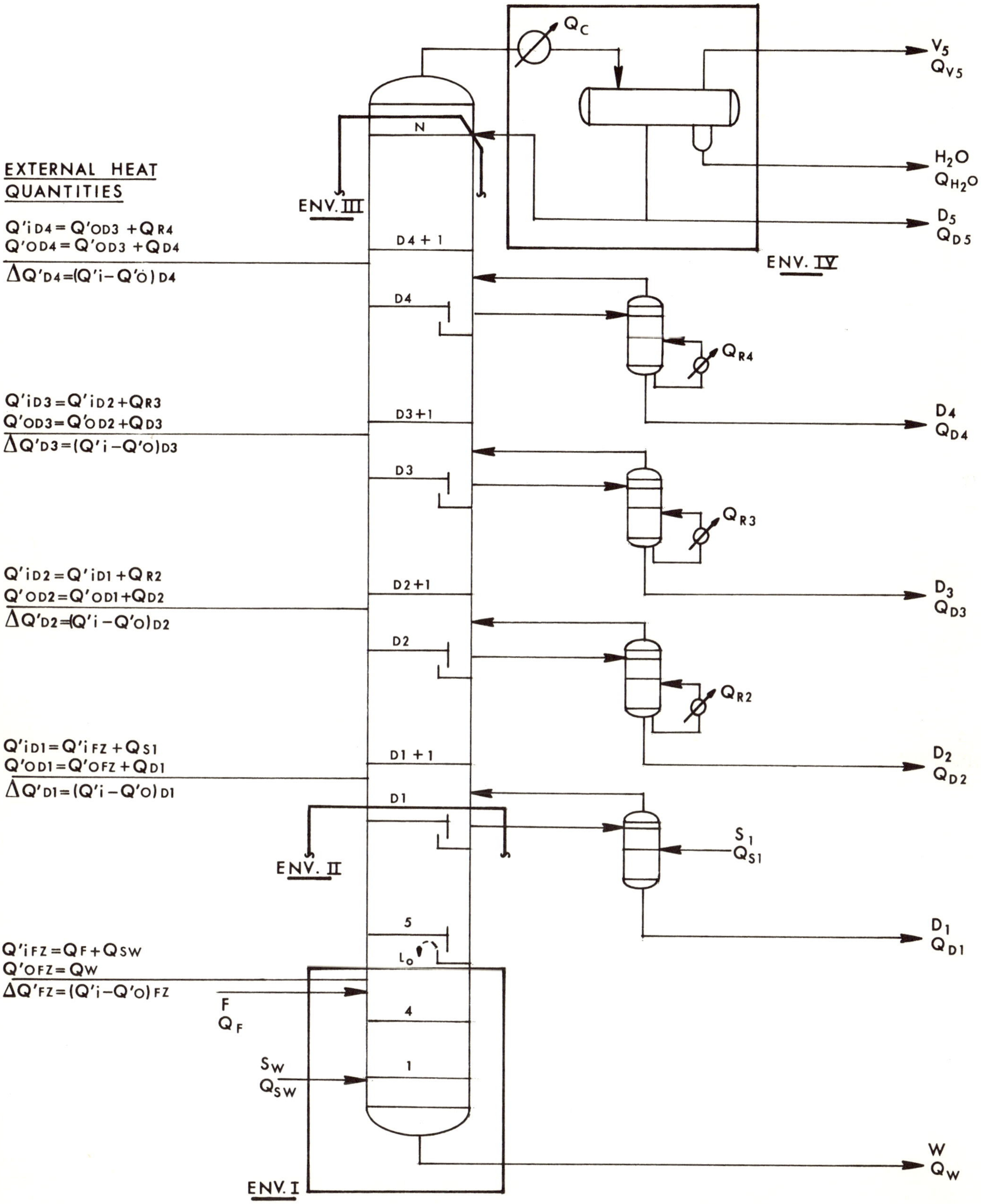

Figure 2.17. Heat and material balance summary—atmospheric tower, Type U.

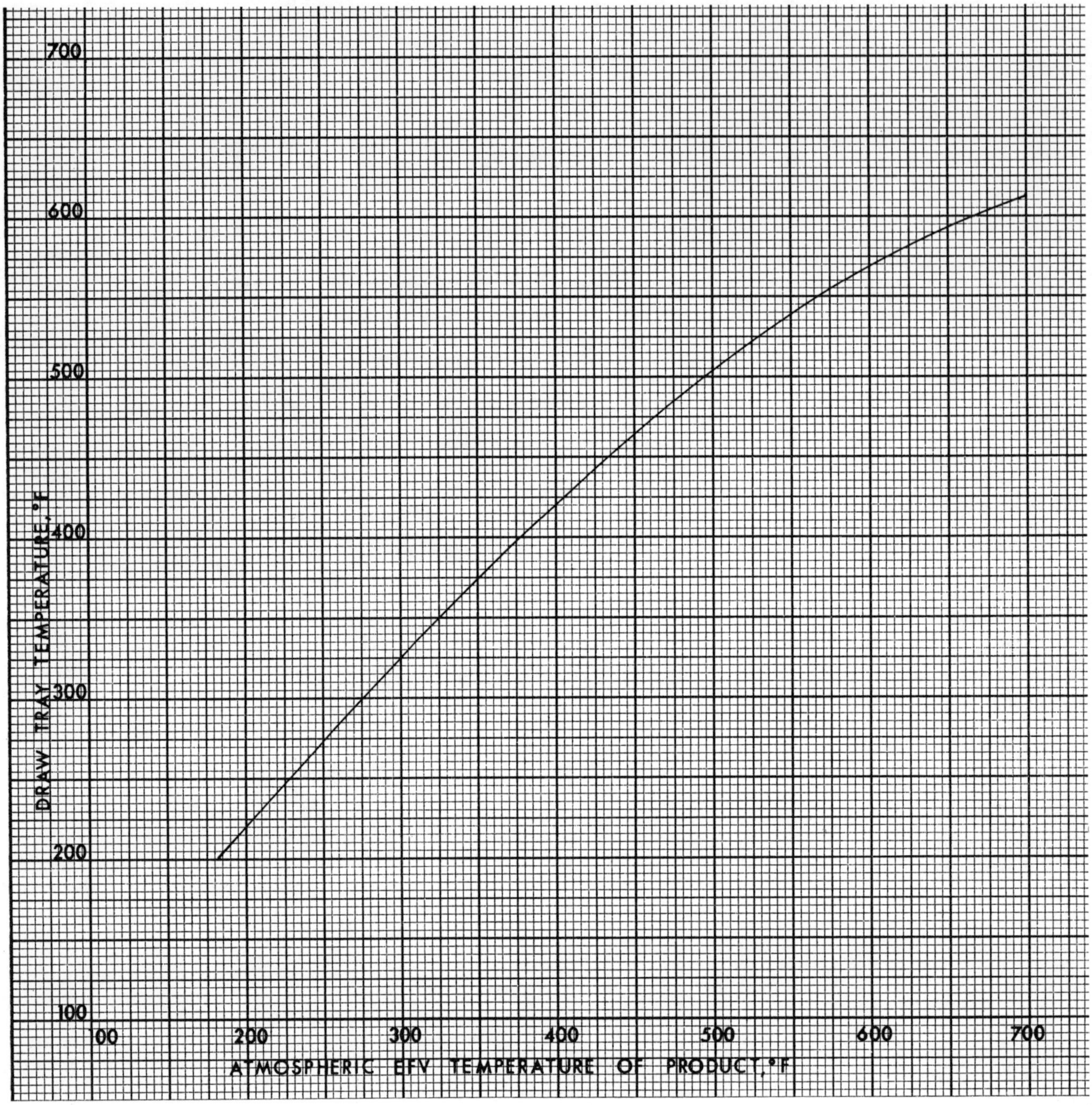

Figure 2.18. Draw tray temperatures—Type U tower.

seen later, it simplifies the calculations for the Type A system to a very great degree.

This section outlines procedures for calculating product draw tray temperatures at all points in the tower and for making an overall heat balance around the system. The method is based upon assuming a draw tray temperature and then calculating the internal reflux required by the system's heat balance. This internal reflux to the draw tray defines the hydrocarbon product partial pressure in the vapor above the tray. Converting the 14.7 psia bubble point of the unstripped liquid to this partial pressure gives a temperature which must check the assumed value. The top tray temperature is calculated by the same procedure, except that the reference temperature is the 14.7 psia dew point of the gross overhead product.

Preliminary Steps

At this point, it is assumed that all the calculations outlined in the previous sections of this chapter have been performed and that the following have been fully defined.

1. Complete hydrocarbon material balance for feed and products.
2. Steam rates to stripping sections and steam distribution between overhead distillate vapor and liquid.
3. Hydrocarbon material balances around product strippers.
4. Atmospheric EFV temperatures for products corresponding to the estimated stripout for each product.
5. Draw tray locations, number of trays in each section and total number of trays in the tower.
6. Heat input to the base section of the tower from feed and bottoms-stripping steam, heat outflow in the bottoms liquid and external heat quantities at the flash zone. This bottoms–section heat balance is shown as Envelope I on Figure 2.17.

All these items must be completed before proceeding further with calculations.

Estimate of Tower Operating Conditions

Draw tray temperatures are estimated from the correlation of Figure 2.18. The following form the basis for this chart.

1. Flash-zone pressure = 24.7 psia.
2. Overflash = 2.0 volume percent of feed.
3. Reduced crude and lowest sidestream are steam stripped at 10 pounds steam per barrel of product measured as 60 degree F liquid.
4. All other sidestreams are reboiled equivalent to the steam stripping at 10 pounds per barrel.
5. For sidestream products, use the estimated bubble point of the unstripped liquid from draw tray. For the total overhead product, use calculated dew point.

Note carefully the restrictions which apply and the indicated temperature variations which will occur as the process conditions differ from the stipulated bases.

1. An increase in flash-zone pressure will increase draw tray temperatures.
2. An increase in overflash will slightly decrease draw tray temperatures of the second sidestream product and all others above.
3. An increased stripping steam rate will decrease product draw tray temperatures due to the reduced hydrocarbon partial pressures.
4. Use of stripping steam in all product strippers rather than in only the first sidestream stripper will decrease draw tray temperatures of the second sidestream and all others above.

Having estimated draw tray temperatures and having set the operating pressures at key points in previous work, plot a temperature and pressure profile for the tower by assuming linear change between draw trays.

Some investigators have stated that 40 degree F temperature rises are experienced between the draw tray and the first and second trays below the draw and that a 15 degree F drop occurs between the draw tray and the one above. This does not seem logical in many cases, particularly in the lower sections of the tower. It seems particularly suspicious in the light of stating that heat removal is more effective on one of two adjacent trays when both trays are handling markedly similar fluids. Unless specific process information to the contrary is available, the assumption of linear temperature drops between draw trays is recommended since it results in greater reflux requirements and tower size. Thus, any error inherent in the assumption is on the safe side.

First Sidestream Product Draw Tray

The heat and material balance relationships at this section of the tower are determined by making a balance around Envelope II as shown on Figure 2.17. An expanded view of this is illustrated by Figure 2.19 which shows the equations used in making the calculations. These equations are to be used in the following sequence.

1. Calculate the reflux heat, Q_R at Tray D1. Reflux heat is defined as the apparent heat imbalance between external heat quantities at the point in question in the tower. These external heat quantities are denoted as Q with appropriate subscripts to signify their location. External heat input quantities are defined as the heat contained in the feed plus all heat to the system at product strippers either directly as steam or indirectly through reboilers. External heat output quantities at a given tray are defined as the heat contained in liquid products leaving the system from points lower in the tower, the heat contained in the internal vapors of products plus steam and the heat contained by a product liquid flowing to the sidestream stripper. If the tray is not a sidestream draw tray, this latter quantity does not enter into the heat balance.

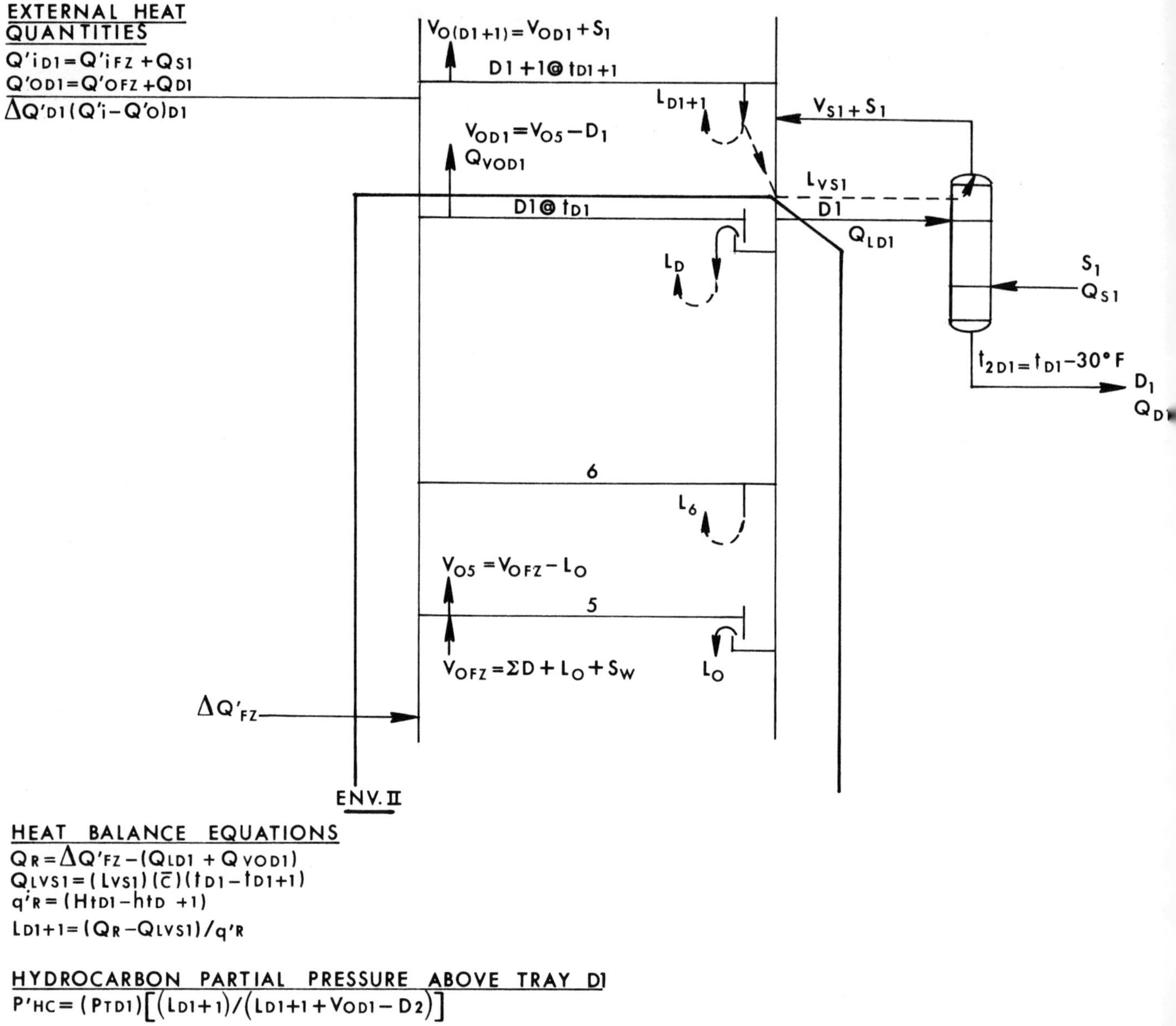

Figure 2.19. Heat and material balance quantities at first sidestream product draw tray—type U tower.

2. The hydrocarbon which is to be revaporized in the product stripper falls to the draw tray as part of the internal reflux from Tray (D1 + 1) rather than rising to the tray as part of the product vapors. In passing across the draw tray, this liquid absorbs a small amount of the reflux heat. This heat absorption is calculated as Q_{LSV1}. Note from the example calculations that this heat quantity is quite small when compared to the reflux heat and could be neglected with no discernible loss of accuracy. It is included here for the sake of total definition of tower flows and heat quantities.

3. Calculate the heat removal capability of the reflux available to the tray as q'_R.

4. Calculate the internal reflux required to absorb the excess heat at Tray D1 as L_{D1+1} and convert it to moles per hour.

5. Calculate the mole fraction of hydrocarbon product vapor in the total vapor leaving the draw tray, neglecting the presence of the product to be removed on the next draw tray up in the tower. The reasoning behind this is that the product vapors which are to be withdrawn from draw trays above the next draw tray are

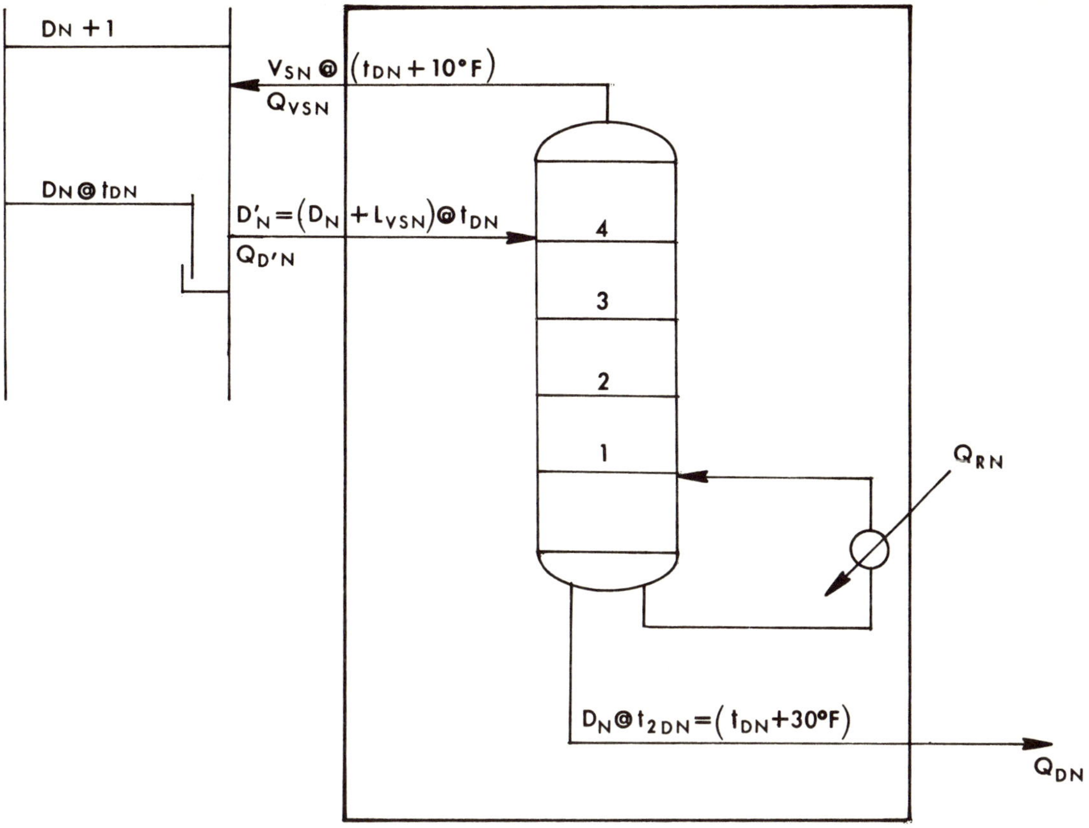

$$Q_{RN} = (Q_{VSN} + Q_{DN}) - Q_{D'N}$$

Figure 2.20. Heat and material balance quantities around a reboiled stripper.

Table 2.4

Product Vapor to be Neglected In Partial-Pressure Calculations	
Draw Tray	Product Vapor
D2	D3
D3	D4
D4	D5 – V5*
*To be considered as an inert gas.	

above their critical temperatures when they are at the temperature of the tray under study. Thus, like the steam, they behave as noncondensibles and lower the boiling point of the product in accordance with Dalton's Law of partial pressures. Since the product vapor which is to be removed at the next higher draw tray will be near its critical temperature as it leaves this lower draw tray, it is assumed to have no effect on the partial pressure, and for this reason, its presence is neglected.

6. Calculate the hydrocarbon product partial pressure in the total vapor leaving the draw tray and convert the atmospheric bubble point of the unstripped liquid product on the tray to this partial pressure. The vapor pressure data in Maxwell, Section 4, is recommended for this conversion. If this temperature does not check the value which was originally assumed, repeat the procedure for a new assumed temperature.
7. Calculate and tabulate the external heat quantities to the base of Tray (D1 + 1). This will include the heat input to the system by the steam to the product stripper and the heat output from the system in the product liquid D1. For steam strippers in atmospheric towers, a temperature drop of 30 degrees F for the liquid from the draw tray to the stripper outlet is assumed arbitrarily.

For reboiled strippers, a liquid temperature rise of 30 degrees F is assumed from the draw tray to the stripper outlet. The temperature of the stripout returning to the atmospheric tower is assumed to be 10 de-

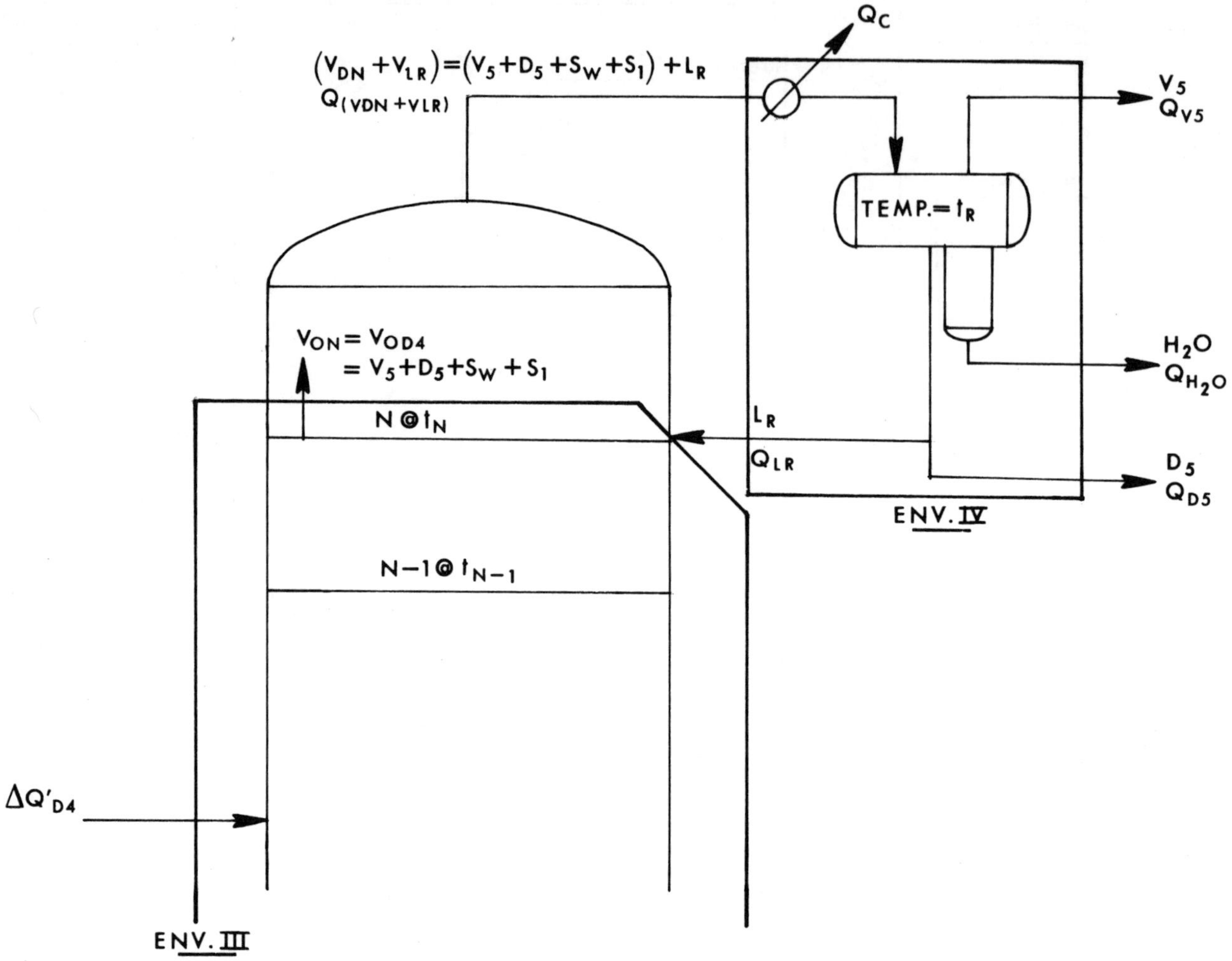

EQUATIONS FOR ENVELOPE III

(1) HEAT BALANCE

$Q_R = \Delta Q'_{D4} - Q_{VN}$

$(q'_R)_{LR} = (H_{tN} - h_{tR})$

$L_R = Q_R / q'_R$

(2) HYDROCARBON PARTIAL PRESSURE ABOVE TRAY N

$P'_{HC} = (P_{TN})\left[(L_R + D_5)/(L_R + D_5 + S_W + S_1)\right]$

(3) INDUCED REFLUX ON TRAY N

$(R_i)_N = (L_R)\left[(h_{tN} - h_{tR})/(H_{t_{N-1}} - h_{t_N})\right]$

EQUATIONS FOR ENVELOPE IV

(1) CONDENSER HEAT BALANCE

$Q_R = Q_{(VON + VLR)} - (Q_{V5} + Q_{H_2O} + Q_{D5} + Q_{LR})$

(2) OVERALL SYSTEM HEAT BALANCE

$Q_R = \Delta Q'_{D4} - (Q_{V5} + Q_{H_2O} + Q_{D5})$

Figure 2.21. Heat and material balance quantities at top tray and condenser.

grees F hotter than the temperature of the draw tray. The trend in recent years has been to specify reboiled strippers wherever thermally possible. This is an economical approach since steam increases the size of the tower and the overhead condenser and, also, since steam requires greater capacities in facilities for boiler feedwater preparation and for treating the sour water effluent from the condenser. The procedure for calculating the heat duty of the reboiler is illustrated by Figure 2.20.

8. Calculate and tabulate the vapor and liquid quantities to the base of Tray (D1 + 1).

Other Sidestream Product Draw Trays

The remaining sidestream draw trays are calculated by the same procedure as that outlined in the previous step. Remember that, in making partial pressure calculations, the presence of the next higher product vapor in the total vapor leaving the draw tray must be neglected. This principle is summarized in Table 2.4.

Top Tray Calculations

The heat and material balance relationships at the top tray are determined by making a balance around Envelope III as shown on Figure 2.17. Figure 2.21 shows an expanded view of this section of the tower and gives the equations to be used in making the calculations. These equations are to be used in the following sequence.

1. Calculate the reflux heat above the top tray.
2. Calculate the heat removal capability of the available reflux.
3. Calculate the amount of reflux from the condenser which is required to absorb the excess heat at Tray N as L_R and convert it to moles per hour.
4. Calculate the mole fraction of hydrocarbon product vapor in the total vapor leaving the top tray, neglecting the hydrocarbon portion of the distillate vapor product.
5. Calculate the hydrocarbon partial pressure of the overhead distillate liquid plus reflux in the total vapor leaving Tray N. Convert the atmospheric dew point of the distillate liquid product to this partial pressure. If this calculated temperature does not check the value originally assumed, assume a new temperature and repeat the procedure.
6. Calculate the reflux induced on the top tray. Induced reflux is the amount of vapor from Tray (N - 1) which enters and is condensed on Tray N for the purpose of raising the reflux liquid from its temperature leaving the reflux drum to that of the top tray.
7. Calculate and tabulate the vapor and liquid quantities at Tray N.

Condenser Calculations and Overall Heat Balance

The condenser duty is calculated by making a heat balance around Envelope IV in Figure 2.17. An expanded view of this calculation is given in Figure 2.21. Note that, in Figure 2.21, Q_c is also calculated by an overall system heat balance. The two values should check to within 2 percent of the absolute value of Q_c. If not, a serious error has been made somewhere in the calculations, and the work must be repeated until a satisfactory check is obtained.

Fractionation Capability of the System

Having completed the heat and material balance calculations around the atmospheric tower and its auxiliaries, it must now be determined whether or not the fractionation specifications can be met with the resulting configuration of trays and reflux. The reflux which exists in the tower is determined principally by the definition of the material balance and is influenced relatively little by draw tray locations. Because of tray pressure drop, a slight variation in calculated temperatures will result by altering draw tray locations, but, unless the change in configuration is drastic, say by four trays or more, the effect on reflux will be slight, certainly by less than 10 percent. Thus, for all practical purposes, reflux in a Type U system may be defined as constant for a given material balance. Referring back to figures 2.6 and 2.7, the technique for analyzing fractionation capability is outlined by the following procedure.

1. Since Packie's reflux ratio is defined in terms of liquid falling from a draw tray, calculate the reflux from all draw trays and from the top tray by making a heat balance around the system at all trays below these just mentioned. These temperatures on the trays below draw trays are arrived at by the assumption of linear temperature variation between draw trays. Calculate these reflux flows as both moles and volumes.
2. Calculate the reflux ratio as the volume of reflux from the draw tray per volume of total product vapors entering the draw tray. The separation factor, F, is the product of this reflux ratio and the number of actual trays in the section.
3. Calculate the difference between the ASTM 50 volume percent temperatures of the product and that of the total remaining lighter products.
4. From Packie's curves, read the ASTM (5-95) Gap for the various separations and compare them with the required values.

For a Type U system using conventional numbers of trays between draw trays and requiring normal separations, there will always be an excess of "trays x reflux". This leads to the conclusion that, by removing heat from the system at points relatively low in the tower, one can reduce reflux and thus reduce tower size while still being able to meet fractionation specifications. This leads into the study of the two types of towers employing heat removal systems.

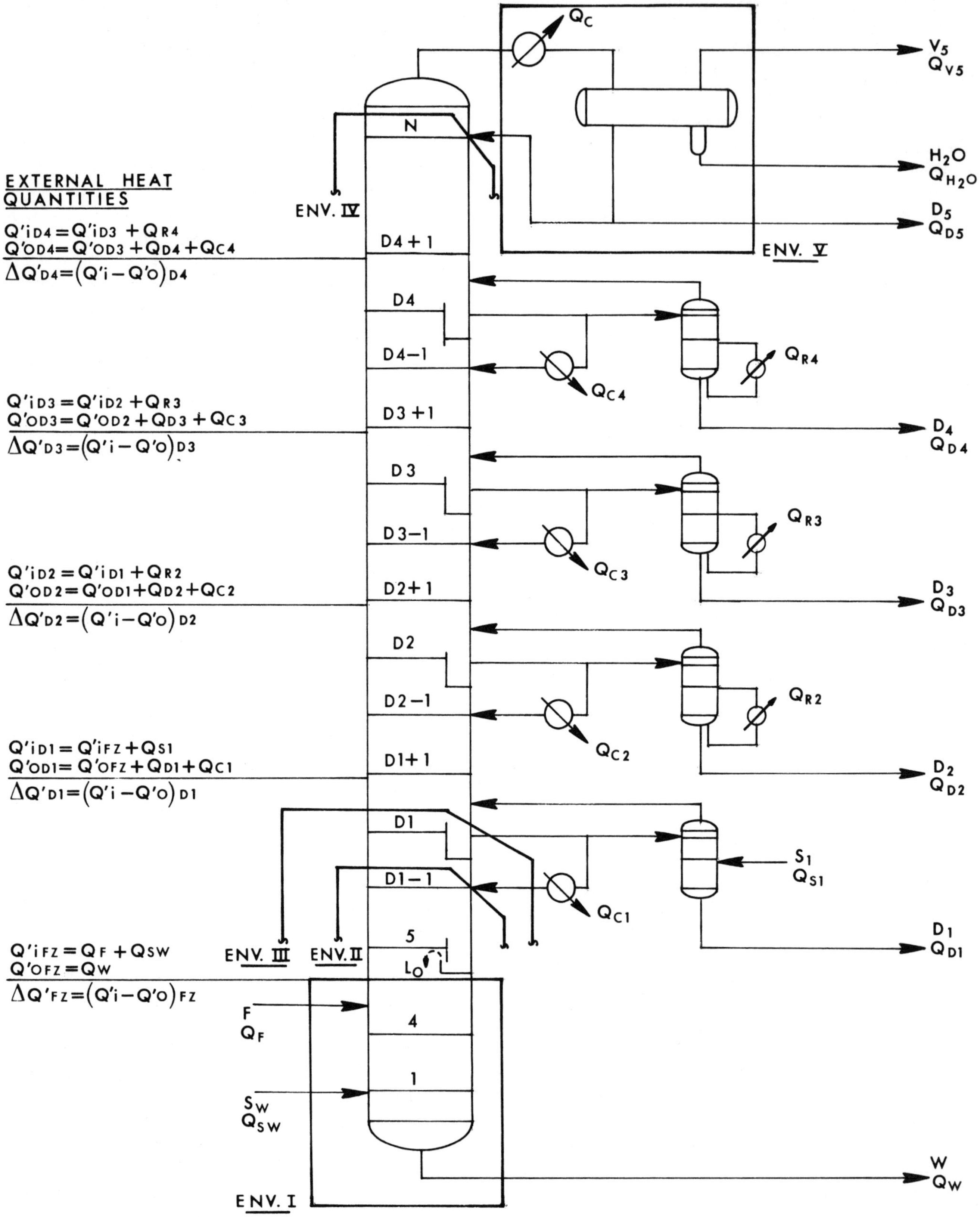

Figure 2.22. Heat and material balance summary—atmospheric tower, Type R.

Another facet of the results of the analysis of fractionation capability is that, in the Type U calculation example, there appears to be a slight imbalance between the various sections. While there is more than enough "trays x reflux" for the required separations, the conditions in the heavy naphtha-light distillate section has considerably less excess "trays x reflux" than do all the other sections. This implies an incorrect choice of draw tray location.

Summary of Vapor-Liquid Traffic

Tabulate the flows of vapor and liquid at all key trays in the tower as moles per hour. Plot these values versus tray number as vapor from tray and liquid to tray. This plot will be of great assistance in tower sizing calculations.

Heat and Material Balance Calculations for Type R Towers

A complete Type R tower is shown in Figure 2.22. This drawing illustrates the basic process and its essential auxiliaries as well as the external heat and material balance quantities. In this type of installation, sidestream draw trays are total draws, i.e., the total liquid leaving the draw tray—sidestream product plus reflux to the tray below—is withdrawn from the tower. The reflux is pumped back to the tower after cooling rather than overflowing internally from the draw tray to the tray below as in the case of Type U and Type A systems. Figure 2.22 will be the basis for discussing the heat and material balance calculations in this section.

The method of splitting the draw tray exit liquid upstream of the product stripper is the preferred arrangement from the viewpoint of operating economy. The alternate processing arrangement allows the total liquid to flow to the stripper. The stripped liquid is cooled and then split into product and pumpback reflux. While the former arrangement requires two sets of pumps and heat exchangers as opposed to only one set required by the latter, the additional operating cost for stripping the reflux more than offsets the differences in capital investment. In most cases, the pumpback reflux will be a significantly greater volume than the product. In view of this, stripping of reflux in addition to the product will require total heat inputs to the strippers considerably greater than that required for stripping only products. If steam stripping were used rather than reboiling, this would necessitate a significantly greater atmospheric tower diameter as well as larger facilities for boiler feedwater treating and for handling of foul condensate from the tower overhead.

This section outlines procedures for calculating product draw tray temperatures at all points in the system and for making an overall heat balance around the tower. The method is based upon assuming a draw tray temperature and then calculating the internal reflux required by the system's heat balance. This internal reflux to the draw tray defines the hydrocarbon product partial pressure in the vapor above the tray. Converting the 14.7 psia bubble point of the unstripped liquid to this partial pressure gives a temperature which must check the assumed value. The top tray temperature is calculated by the same procedure, except that the reference temperature is the 14.7 psia dew point of the gross overhead product.

Preliminary Steps

At this point, it is assumed that the following have been fully defined.

1. Complete hydrocarbon material balance for feed and products.
2. Steam rates to stripping sections and steam distribution between overhead distillate vapor and liquid.
3. Hydrocarbon material balances around product strippers.
4. Atmospheric EFV temperatures for products corresponding to the estimated stripout for each product.
5. Draw tray locations, number of trays in each section and total number of trays in the tower.
6. Heat input to the base section of the tower from feed and bottoms stripping steam, heat outflow in the bottoms liquid and external heat quantities at the flash zone. This bottoms-section heat balance is shown on Envelope I in Figure 2.22.

All these items must be completed before proceeding further with calculations.

Estimate of Tower Operating Conditions

Draw tray temperatures are estimated from the correlation of Figure 2.23. The following form the basis for this chart.

1. Flash zone pressure = 24.7 psia.
2. Overflash = 2.0 volume percent of feed.
3. Reduced crude and lowest sidestream are steam stripped at 10 pounds steam per barrel of product, measured as 60 degree F liquid.
4. All other sidestreams are reboiled equivalent to steam stripping at 10 pounds per barrel.
5. Cooled, unstripped pumpback reflux is used at each sidestream draw tray.
6. For sidestream products, use the estimated bubble point of unstripped liquid from the draw trays. For overhead product, use the calculated dew point.

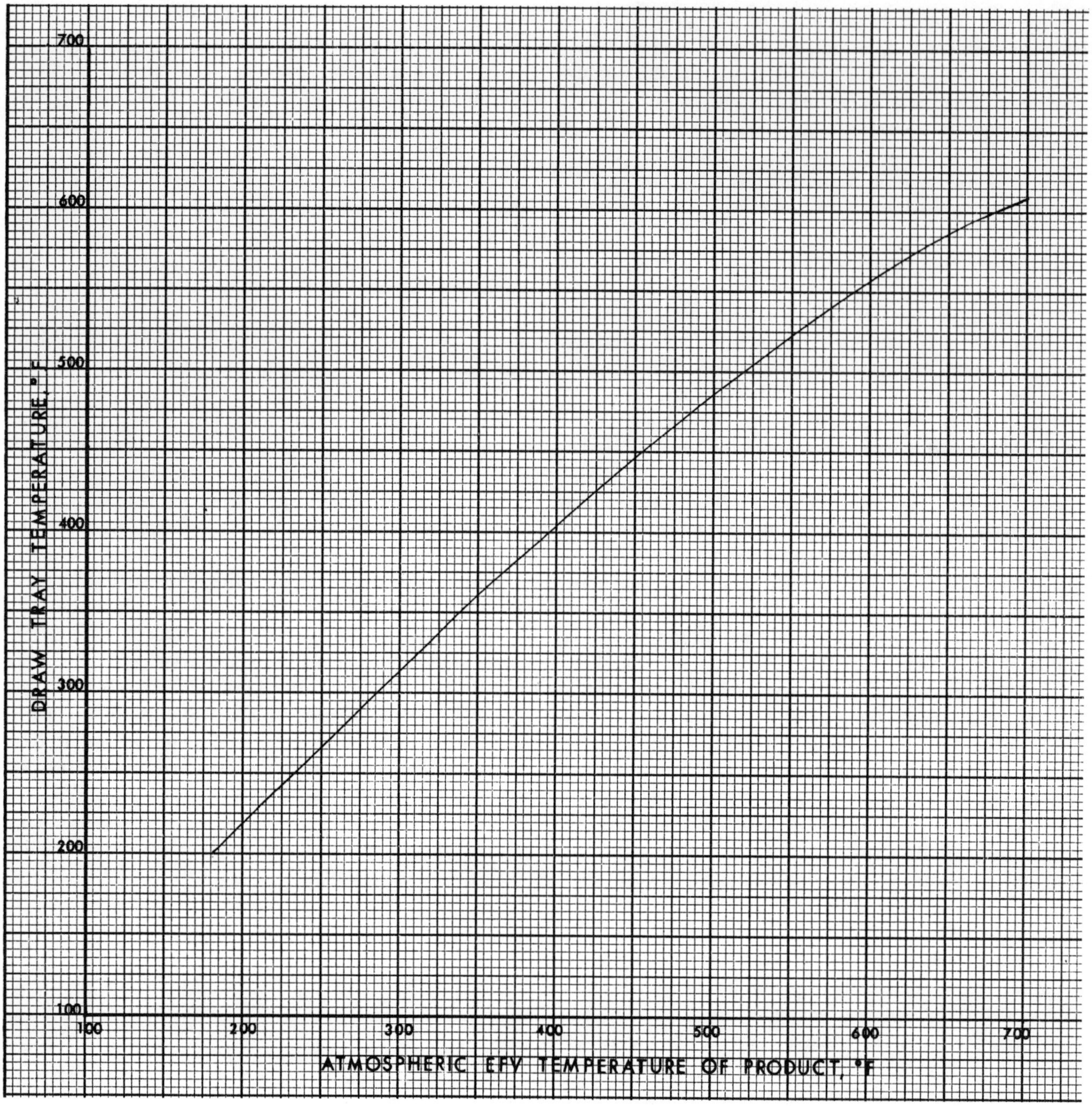

Figure 2.23. Draw/Tray temperatures—Type R tower.

Note carefully the restrictions which apply and the indicated temperature variations which will occur as the process conditions differ from the stipulated bases.

1. An increase in flash zone pressure will increase draw tray temperatures.
2. An increase in overflash will slightly decrease draw tray temperatures of the second sidestream product and all others above.
3. An increased stripping steam rate will decrease product draw tray temperatures due to the reduced hydrocarbon partial pressures.
4. Use of stripping steam in all product strippers rather than in only the first sidestream stripper will decrease draw tray temperatures of the second sidestream and all others above.
5. Stripping of reflux in addition to stripping the product will increase heat input to the system. If this is by

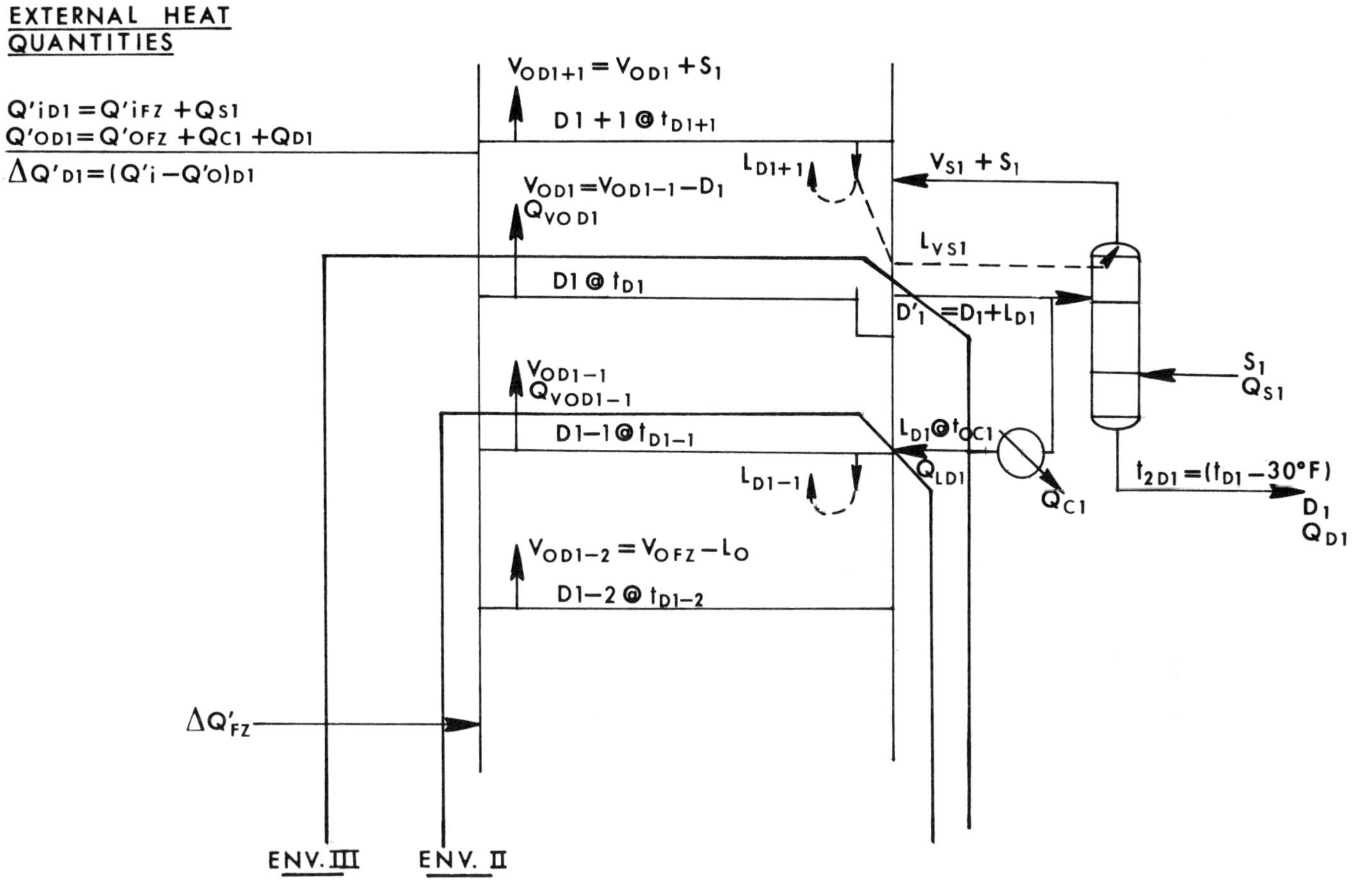

HEAT BALANCE EQUATIONS

(1) ABOVE TRAY (D1−1)—ENVELOPE II

$$Q_R = (\Delta Q'_{FZ} - Q_{VOD1-1})$$
$$q'_R = (H_{tD1-1} - h_{toC1})$$
$$L_{D1} = Q_R / q'_R$$

(2) ABOVE TRAY D2—ENVELOPE III

$$Q_R = [(\Delta Q'_{FZ} + Q_{LD1}) - (Q_{VOD1} + Q_{D'1})]$$
$$Q_{LVS1} = (L_{VS1})(\bar{c})(t_{D1} - t_{D1+1})$$
$$q'_R = (H_{tD1} - h_{tD1+1})$$
$$L_{D1+1} = (Q_R - Q_{LVS1}) / q'_R$$

HYDROCARBON PARTIAL PRESSURE ABOVE TRAY D1

$$P'_{HC} = (P_{TD1})[(L_{D1+1}) / (L_{D1+1} + V_{OD1} - D2)]$$

INDUCED REFLUX ON TRAY (D1−1)

$$R_{iD1-1} = L_{D1}[(h_{tD1-1} - h_{toC1}) / (H_{tD1-2} - h_{tD1-1})]$$

Figure 2.24. Heat and material balance quantities at first sidestream product draw tray—Type R tower.

reboiling, temperatures at the next draw tray upward and all those higher will be increased. If this is by steam stripping, however, these temperatures will be lowered due to the reduction of hydrocarbon partial pressures.

Having estimated draw tray temperatures and having set the operating pressure profile in previous work, plot a temperature and pressure profile for the tower by assuming linear change between draw trays.

Estimating the temperature of the cooled pumpback reflux requires a good deal of experience with the crude oil upon which the design is being based, and, unfortunately, this is usually only available in an operations-analysis situation. In design work, this temperature must be estimated, and later, after the heat duties and temperature levels of all the reflux coolers have been determined, it must be verified that the crude oil in being preheated is capable of absorbing the heat from the pumpback reflux streams. Optimizing the crude preheat-tower cooling heat-exchange train is the heart of crude unit design, and each case must be studied on an individual basis in order to arrive at the most economical processing scheme. It should be remembered that most designs will be based on several different operations, and this will exert a far greater influence on the design of the heat exchange equipment than will the assumed temperature levels alone. Since a large portion of the reflux heat removal is accomplished by the latent heat of vaporization and a relatively small portion by sensible heat, some latitude is available in choosing the cool reflux temperatures. The key point is not to assume a temperature level which is unattainably low. High values can be adjusted downward, but low values will require redoing the work. This subject would be more fittingly treated in a work on process optimization and is mentioned here to alert the reader to possible difficulties which will result from assuming unrealistic cool reflux temperatures.

First Sidestream Product Draw Tray

The heat and material balance relationships at this section of the tower are determined by making two balances which are shown as Envelopes II and III on Figure 2.22. An expanded view of these is illustrated by Figure 2.24, which also gives the equations to be used in making the calculations. These equations are to be used in the following sequence. Note that most of these points have already been encountered in the discussion of the Type U tower. Thus, detailed discussion will be presented here only to explain new concepts.

Balance above Tray (D1 – 1)–Envelope II

1. Calculate the reflux heat at Tray D1 – 1.
2. Calculate the heat removal capability of the cooled pumpback reflux.
3. Calculate the amount of cool pumpback reflux which is required to absorb the reflux heat at Tray (D1 – 1). Calculate the heat content of this stream as it reenters the tower.

Balance above Tray D1–Envelope III

1. Calculate the reflux heat at Tray D1. The liquid leaving this draw tray is the sum of the product, D1, and the pumpback reflux, L_{D1}. The liquid which is revaporized in the product stripper falls to the draw tray as a small part of the reflux from Tray (D1 + 1) and, thus, for the purpose of this calculation, is not considered part of the exit liquid. Also note that, if the process is to be designed for stripping of the pumpback reflux, the stripout liquid falling to Tray D1 will be accordingly greater.
2. Calculate the heat absorbed by the stripout liquid in passing across the draw tray.
3. Calculate the heat removal capability of the internal reflux falling from Tray (D1 + 1).
4. Calculate the internal reflux from Tray (D1 + 1) which is required to absorb the excess heat at Tray D1.
5. Calculate the mole fraction of product vapors in the total vapors leaving the draw tray, remembering to neglect the presence in the vapor of the product which is to be withdrawn on the next draw tray up in the tower.
6. Calculate the hydrocarbon partial pressure of product in the total vapor leaving the draw tray. Convert the atmospheric bubble point of the unstripped liquid on the draw tray to this partial pressure. If this temperature does not check the value assumed earlier, repeat the procedure for a new assumed temperature. If recalculation is required, it is usually not necessary to revise the assumed temperature of the cooled pumpback reflux unless, for some reason, a gross error was made in assuming the draw tray temperature.
7. Calculate the heat effect at the sidestream stripper.
8. Calculate the reflux induced on Tray (D1 – 1) as the amount of vapor from Tray (D1 – 2) which enters and is condensed on Tray (D1 – 1) for the purpose of converting the subcooled pumpback reflux liquid to its bubble point.
9. Calculate and tabulate the vapor and liquid quantities to the base of Tray (D1 + 1).
10. Calculate and tabulate the external heat quantities to the base of Tray (D1 + 1).

Other Sidestream Product Draw Trays

The remaining sidestream draw trays are calculated by the same procedure as that outlined in the previous step. Remember that, in making partial pressure calculations, the presence of the next higher product vapor in the total vapor leaving the draw tray must be neglected.

Top Tray Calculations

The heat and material balance relationships at the top tray are determined by making a balance around Envelope IV as shown in Figure 2.22. The detailed flows around this section of the tower and the equations to be used in the calculations are the same as those given in Figure 2.21.

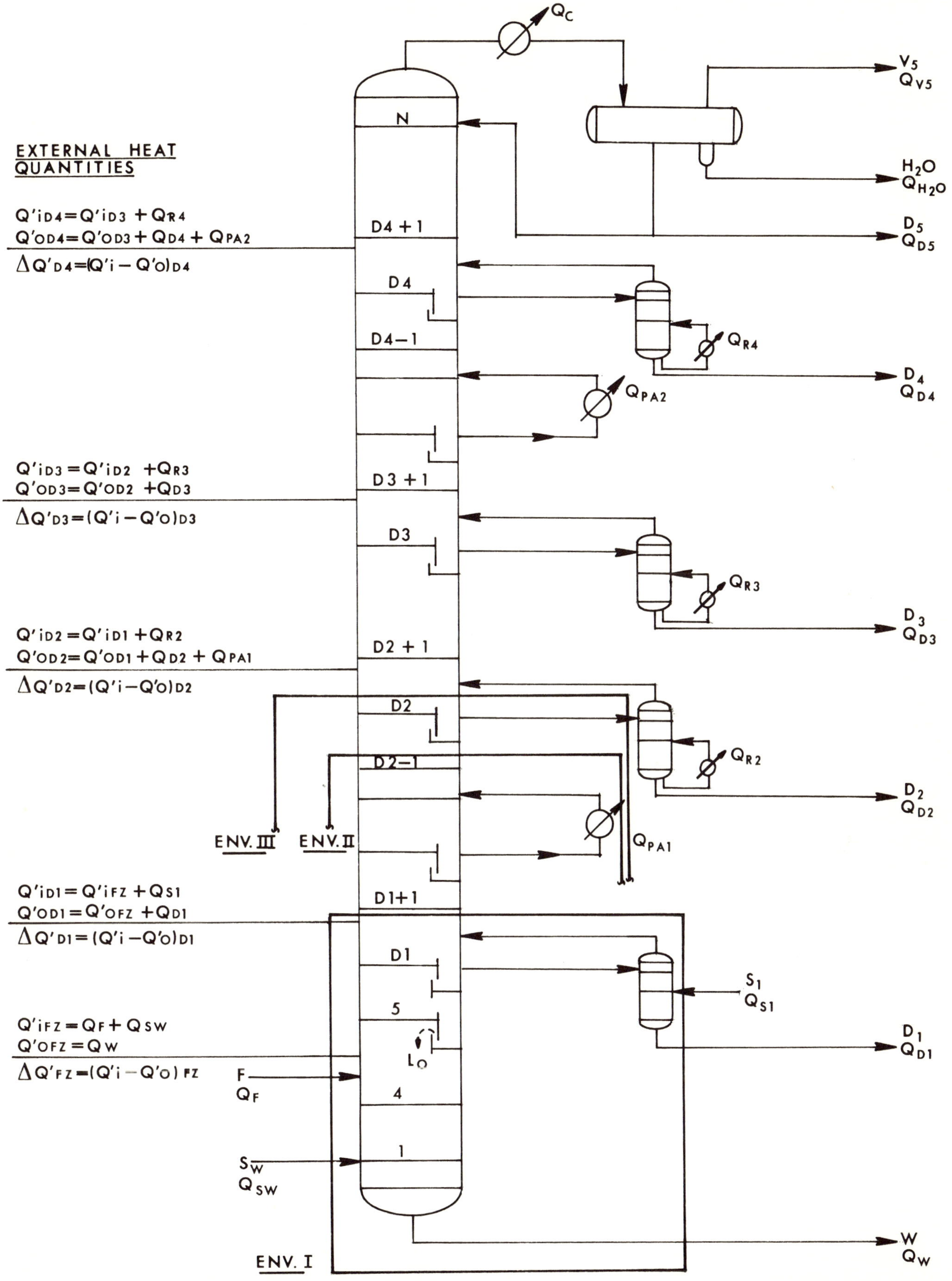

Figure 2.25. Heat and material balance summary—atmospheric tower, Type A with two pumparound circuits.

Condenser Calculations and Overall Heat Balance

The condenser duty is calculated by making a heat balance around Envelope V on Figure 2.22. This must be checked against an overall system heat balance. The two values must check to within 2 percent of the absolute value of the condenser duty.

Fractionation Capability of the System

This subject has been discussed in detail earlier in this work. The only new point to be considered here is the definition of reflux from the draw tray. For purposes of Packie's analysis, reflux is defined as the volume of liquid falling from the tray below the draw tray. In terms of the first side stream draw tray, this is calculated by making a heat balance above Tray (D1 – 2) and then calculating the internal reflux from Tray (D1 – 1) which is required to absorb the excess heat.

Summary of Vapor-Liquid Traffic

Tabulate the flows of vapor and liquid at all key trays in the tower as moles per hour. Plot these values versus tray number as vapor from tray and liquid to tray. This plot is of great value in tower design.

Heat and Material Balance Calculations for Type A Towers

A complete Type A Tower is shown in Figure 2.25. This drawing illustrates the basic process and its essential auxiliaries as well as the external heat and material balance quantities. The two pumparound heat removal systems which are shown are for a typical installation and would not necessarily be located in these particular sections of the tower, nor would every tower employ two systems.

As will be seen in the development of the design procedure, there is some flexibility in the choice of draw tray locations and in the design of the pumparound systems. As in the case of the Type R tower, the optimum design of a crude unit utilizes crude oil as the coolant for the various tower heat removals, thus affording feed preheat which is an obvious operating economy. The final task of the designer is to verify that the assumed thermal aspects of the design are practically attainable. Since, by definition, this work is more concerned with the principles of petroleum fractionation rather than those of process optimization, the following procedures are written accordingly. This method may not always result in an optimum design after completion of the first calculation. However, the optimum can be found rather easily in subsequent iterations.

Preliminary Steps

This procedure is based on the assumption that a complete heat and material balance has been calculated for the process as a Type U tower. This analysis contains the following things.

1. A complete heat and material balance.
2. Complete temperature and pressure profile.
3. Draw tray locations and total number of trays.
4. Reflux volumetric flow rates from all draw trays and the top tray.

Internal Reflux Considerations

It is an inherent property of a Type U tower and a given material balance that, since internal reflux flows are at their maximum values, column temperatures will also be at their maximum levels. Any modification to the system which removes tower heat will lower tower temperatures.

Looking back to Packie's work, it is known that for a given separation requirement expressed as (5-95) Gap, a given material balance expressed in terms of ASTM 50 volume percent temperature difference and a given number of trays in the separation section, there is a value, called F, which is the product of the number of actual trays and the volumetric reflux ratio in the section. Thus, a minimum allowable reflux falling from draw trays can be calculated. This is not minimum reflux in the sense that infinite plates are required for the separation. It is minimum allowable operating reflux for the specified number of trays and the required separation.

Calculation of Minimum Reflux

1. Based on Packie's analysis, calculate the minimum reflux which must be available from the product draw trays to make the required separation. These calculations must be made for two cases:
   a. Assume no pumparound heat removal in the section. Thus, the number of trays to be used in Packie's F-Factor is the actual number of trays in the section.
   b. In each section of the tower, assume a three-tray heat removal system. For fractionation purposes, these three trays are defined as being equivalent to one actual tray. Thus, the number of trays to be used in Packie's F-Factor is two less than the actual number of trays.
2. For purposes of comparison, note the reflux available in the tower as a Type U system.

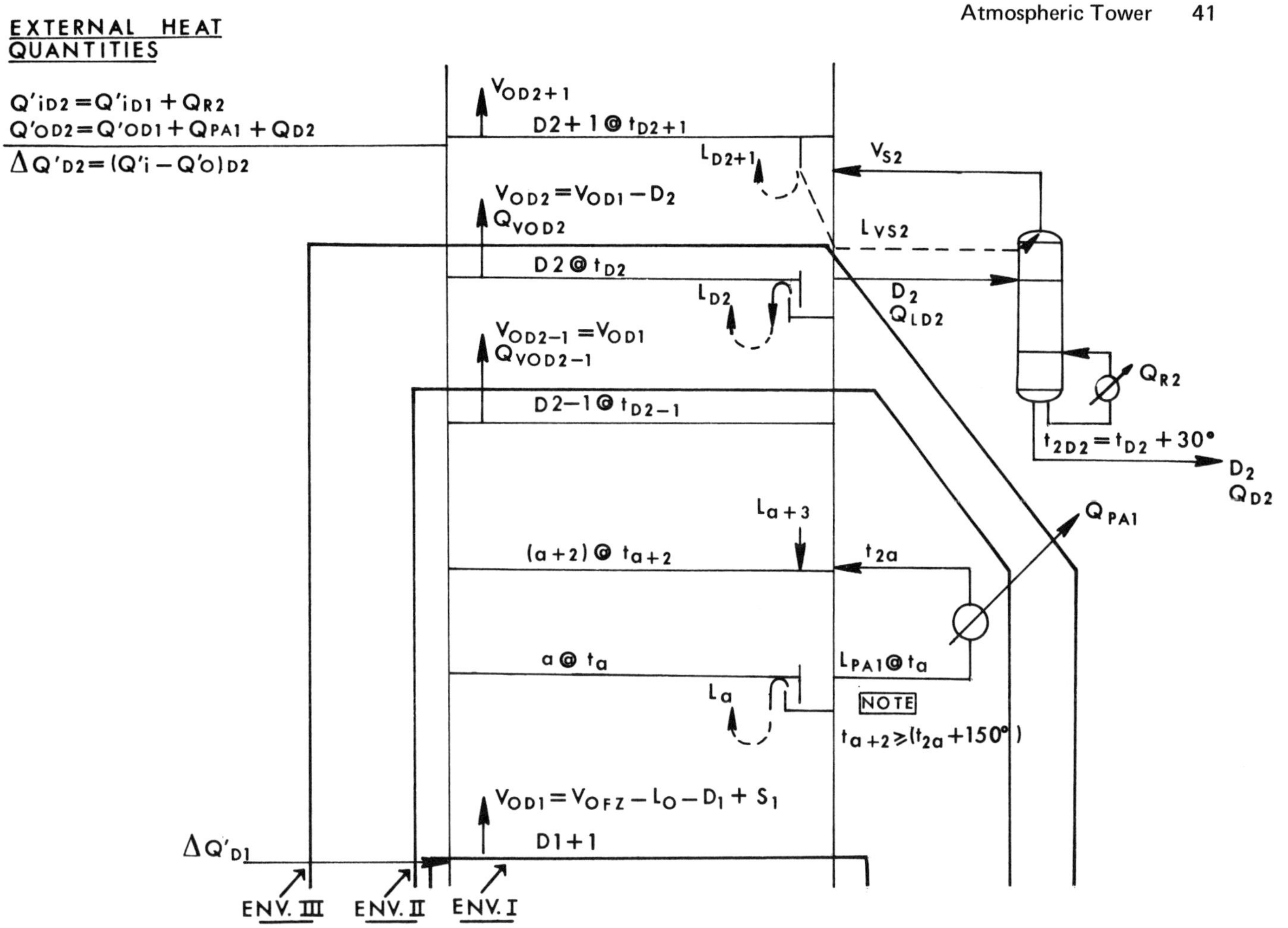

HEAT BALANCE EQUATIONS

(1) ABOVE TRAY (D2–1) – ENVELOPE II

$$\left.\begin{aligned} Q_R &= L_{D2}(H_{tD2-1} - h_{tD2}) \\ Q_R &= \Delta Q'_{D1} - (Q_{VOD2-1} + Q_{PA1}) \end{aligned}\right\} \Longrightarrow Q_{PA1} = \Delta Q'_{D1} - L_{D2}(H_{tD2-1} - h_{tD2}) - Q_{VOD2-1}$$

(2) ABOVE TRAY D – ENVELOPE III

$$Q_R = \Delta Q'_{D1} - (Q_{VOD2} + Q_{LD2} + Q_{PA1})$$

$$Q_{LVS2} = (L_{VS2})(\bar{c})(t_{D2} - t_{D2+1})$$

$$q'_R = (H_{tD2} - h_{tD2+1})$$

$$L_{D2+1} = (Q_R - Q_{LVS2})/q'_R$$

(3) PUMPAROUND CIRCULATION RATE

$$L_{PA1} = Q_{PA1}/(h_{ta} - h_{t2a}) = Q_{PA1}/(\bar{c})(t_a - t_{2a})$$

HYDROCARBON PARTIAL PRESSURE ABOVE TRAY D2

$$P'_{HC} = (P_{TD2})[(L_{D2+1})/(L_{D2+1} + V_{OD2} - D3)]$$

Figure 2.26. Heat and material balance quantities at second sidestream product draw tray—Type A tower.

Estimate of Reflux to Draw Trays at Minimum Reflux Conditions

1. By comparing the reflux available from a Type U operation with minimum reflux requirements for sections having pumparound systems, select the appropriate sections in the column for location of pumparound heat removal systems. Normally, two systems are employed, and they are seldom in adjacent sections.
2. From the Type U calculations, determine the reflux to all draw trays as a function of the total overflow from the draw trays. In all cases, this fraction will be less than 1.0.
3. For all sections, assume that the change in flows across the draw trays will be the same as estimated in Step 2.

4. From Step 3, calculate the internal reflux falling to the draw trays, taking into account the location of pumparound systems.

Estimate of Draw Tray Temperatures at Minimum Reflux Conditions

1. Calculate draw tray temperatures for the case of minimum reflux operation. In making these calculations, remember to neglect the presence in the total vapor leaving the draw tray of the product to be removed on the next draw tray up in the tower.
2. Plot a temperature-pressure profile for the tower. There will be two temperature profiles—one for the maximum values (Type U) and one for the minimum values (minimum reflux).

Estimate of Tower Operating Temperatures for Design

By definition, the reflux from the upper draw tray in a section employing a pumparound system will be the minimum value. Keeping this in mind, estimate the draw tray temperatures and plot this profile on the one developed in the previous section, "Estimate of Draw Tray Temperatures at Minimum Reflux Conditions."

First Sidestream Product Draw Tray

Since there is no pumparound heat removal system in the section between the flash zone and the first sidestream product draw tray, the results of the calculations for this tray will be the same as for a Type U tower. This is shown as Envelope I in Figure 2.25.

Second Sidestream Product Draw Tray

The following discussion assumes a pumparound system between the first and second sidestream product draw trays. The heat and material balance relationships at this section of the tower are determined by making two balances which are shown as Envelopes II and III on Figure 2.25. An expanded view of this section is illustrated by Figure 2.26 which also gives the equations which are to be used in making the calculations. The sequence of calculations is discussed as follows.

Balance above Tray (D2 – 1)—Envelope II

1. Calculate the heat content of the product vapors plus steam leaving Tray (D2 – 1).
2. By definition of a minimum reflux operation, the reflux heat above Tray (D2 – 1) has been preset as the amount which can be absorbed by the predetermined available reflux.
3. The heat which must be removed in the pumparound circuit is now computed from known heat quantities.

Balance above Tray D2—Envelope III

These are conventional calculations above a draw tray and have been covered in detail in earlier portions of this work. The sequence of calculations is as follows.

1. Calculate the reflux heat above Tray D2.
2. Calculate the heat absorbed in passing across the draw tray by the liquid which is to be revaporized in the sidestream product stripper.
3. Calculate the heat removal capability of the reflux available to the draw tray from the tray above.
4. Calculate the reflux required to absorb the excess heat above Tray D2.
5. Calculate the hydrocarbon partial pressure of the product vapor in the total vapor leaving the draw tray, neglecting the presence of the product to be removed at the next higher draw tray.
6. Convert the atmospheric bubble point of the product liquid flowing to the stripper to this partial pressure. There will seldom be any discrepancy between this temperature and the value estimated from the minimum reflux calculations.
7. Calculate the heat effect at the sidestream stripper.
8. Calculate and tabulate the vapor and liquid quantities to the base of Tray (D2 + 1).
9. Calculate and tabulate the external heat quantities to the base of Tray (D2 + 1).
10. Assume that the returning pumparound liquid is 150 degrees F cooler than the tray to which it is being returned. This will provide an adequate temperature difference within the tower for the required heat transfer. Since this also sets the temperature difference for the pumparound liquid, calculate the circulation rate of this stream.

In setting up a design, the location of pumparound systems will usually be predetermined. Thus, a section containing no pumparound is calculated as a conventional Type U draw tray which has been covered earlier.

Top Tray Calculations

Normally, there will be no pumparound system in the section between the highest sidestream product draw tray and the top tray. Thus, it is calculated in the normal manner for a top tray as illustrated by Figure 2.21.

However, if in the actual design calculations, it becomes obvious that there is still a good deal of excess reflux leaving the top tray, another pumparound system may be provided. A side implication of this is that the apportionment of trays between the various sections was incorrect.

Fractionation Capability of the System

In calculating a Type A tower, the fractionation capability is known as a part of the design input. Thus, the calculations required here are routine and have been discussed in full detail in earlier portions of this work.

Summary of Vapor-Liquid Traffic

It is necessary to calculate and to tabulate vapor and liquid quantities at the following locations in addition to those already determined.

1. Liquid to and vapor from the top tray in all pumparound systems.
2. Liquid from and vapor to the bottom tray in all pumparound systems.

References

1. W.L. Nelson, *Petroleum Refinery Engineering,* 4th ed. (New York: McGraw-Hill Book Company, Inc., 1958).
2. J.W. Packie, "Distillation Equipment in the Oil Refining Industry," *AIChE Transactions* 37 (1941), pp. 51-78.
3. W.C. Edmister, *Applied Hydrocarbon Thermodynamics* (Houston: Gulf Publishing Co., 1961).
4. J.B. Maxwell, *Data Book on Hydrocarbons* (Princeton, N.J.: D. van Nostrand Co., 1965).
5. F.W. Winn, "Physical Properties by Nomogram," *Petro leum Refiner* 36, no. 2, (February 1957), p. 157.
6. "NPRA Symposium on Crude Oil Distillation," *Oil and Gas Journal* (March 30, 1970), pp. 93-105.

EXAMPLE HEAT AND MATERIAL BALANCE CALCULATIONS - TYPE U ATMOSPHERIC TOWER

The following calculations will illustrate in detail the procedure for calculating the heat and material balance quantities around a Type U atmospheric crude tower. The design will be based on fractionating dry crude feed into a residuum, four sidedraw distillates, and an overhead yielded as two products, one liquid and one vapor.

1. Process Design Basis

   A. Properties of 36.3° API Crude Oil Feed

      1. Rate = 50,000 barrels per stream day (BPSD).

      2. Light Ends Analysis per 100 bbl. of Oil

Comp.	bbl.	lb.	moles
C_2H_6	0.1	13	0.4
C_3H_8	0.9	177	4.0
iC_4H_{10}	1.4	276	4.7
nC_4H_{10}	2.3	760	13.1
iC_5H_{12}	2.1	459	6.4
nC_5H_{12}	1.8	397	5.5
Total	8.6	2082	34.1

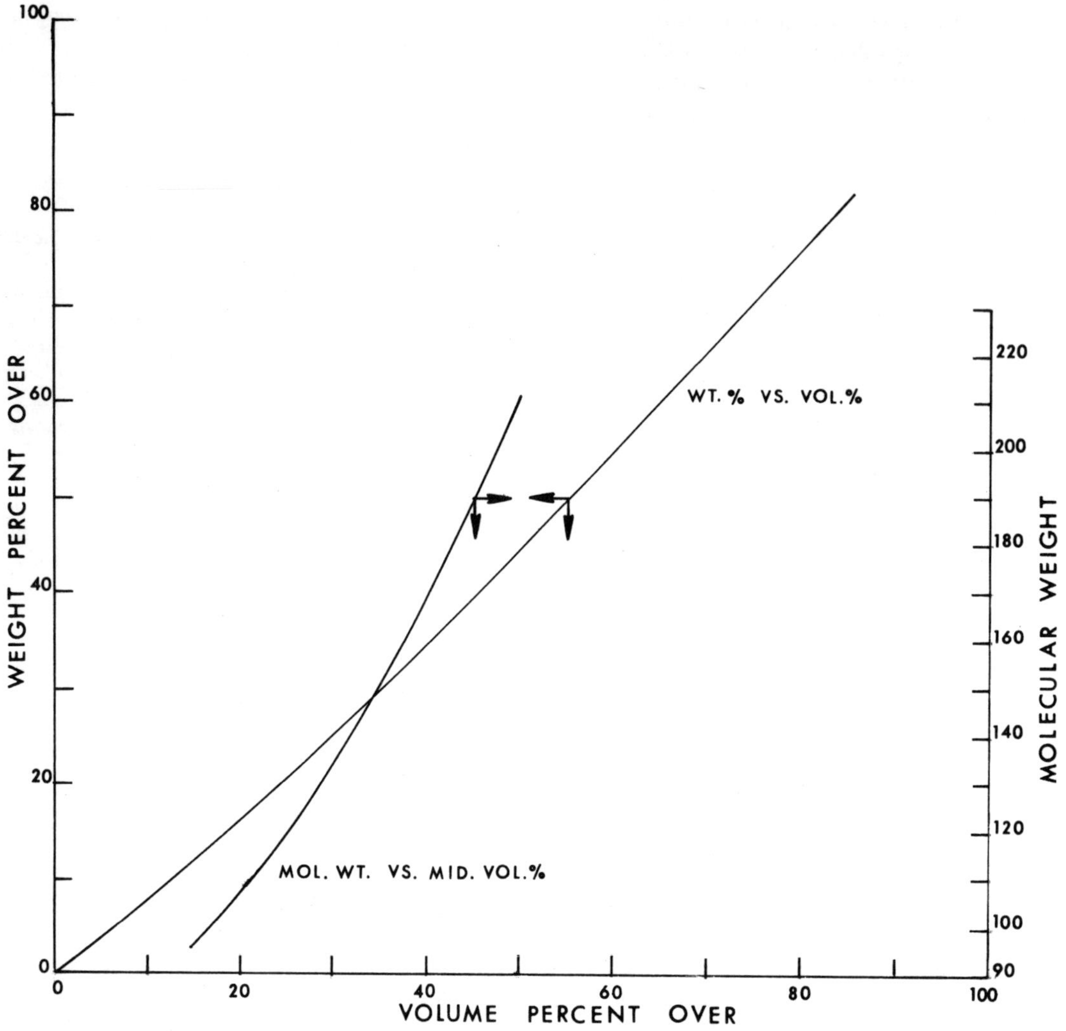

Figure 2.27. 36.3 degree API crude, volume-mole-weight relationships.

3. TBP and Atmospheric EFV Distillations

Vol. %	TBP, °F	EFV, °F
5	62	190
10	144	235
20	255	304
30	344	366
40	437	429
50	531	493
60	623	555
70	717	619
80	819	685
85	897	727

4. The volume-mole-weight relationship for the whole crude is plotted on Fig. 2.27.

B. Product Specifications

This process will be designed to yield, as specification distillates, light naphtha as an overhead and heavy naphtha, light distillate, and heavy distillate as sidedraws. Atmospheric gas oil will also be yielded as a sidedraw liquid as it is available from the total crude vaporization minus (total specification products + overflash). To insure adequate fractionation between streams, product stripping of all sidestreams and the tower bottoms is specified. Atmospheric gas oil and the tower bottoms will be stripped with 65 psia. -500°F steam at a rate of 10 lb./bbl. of net stripped liquid. Heavy distillate and all lighter sidestreams will be reboil-stripped removing an amount of the stripper feed equivalent to steam stripping at a rate of 10 lb./bbl. net stripped liquid.

Criteria for setting the design material balance are as follows.

Product	ASTM End Point, °F	ASTM (5-95), °F
Light Naphtha	275	
Heavy Naphtha	380	+25
Light Distillate	560	+35
Heavy Distillate	650	+10
Atmos. Gas Oil	No Spec.	+ 5
Residuum	-	No Spec.

C. Pressure-Temperature Datum Points

1. The reflux drum will operate at 120°F and 15.0 psia.

2. Pressure drop between the reflux drum and the top tray will be 5.0 psia.

3. Pressure drop between the top tray and the flash zone will be 5.0 psia.

4. Conditions at the outlet of the crude heater will be 700°F and 30 psia.

D. Overflash will be set at 2.0 vol. percent of whole crude feed. It will overflow from the first tray above the flash zone and, together with the feed-flash liquid, will fall to the residuum stripping section.

2. Flash Zone Calculations

The conditions at the furnace outlet have been preset at 700°F and 30 psia. Using Maxwell's vapor pressure-temperature nomogram, this results in the vaporization of 72.4 vol. percent which, from Figure 2.27, is equivalent to 67.8 wt. percent of the feed. Taking a design basis of time as 1 hour,

Crude feed = 2,083 bbl./hr.
= 614,277 lb./hr.

Therefore, leaving the furnace,

Vapor = 1,508 bbl./hr.
= 416,480 lb./hr. @ 47.6° API.
= 2,740 moles/hr. @ 152 mol. wt.

Liquid = 575 bbl./hr.
= 197,747 lb./hr. @ 12.4° API.

The heat input to the process by the feed is now calculated as

$$Q_{FO} = (416{,}480 \times 593) + (197{,}747 \times 456) = 337.145 \text{ MMBTU/hr.}$$

At the entrance to the tower, the total pressure is 25 psia. The temperature and vapor-liquid equilibrium conditions existing there must be determined by a trial-and-error procedure. Assuming a temperature of 695°F, this results in the vaporization of 74.2 vol. percent and 69.8 wt. percent, respectively, of the feed. Thus,

Vapor = 1,545.6 bbl./hr.
= 428,765 lb./hr. @ 46.9° API
= 2,784 mol./hr. @ 154 mol. wt.

Liquid = 537.4 bbl./hr.
= 185,512 lb./hr. @ 11.9° API.

The heat content of the feed is now calculated as

$$Q_{FZI} = (428{,}765 \times 590) + (185{,}512 \times 453) = 337.008 \text{ MMBTU/hr.}$$

This checks Q_{FO} calculated above. In subsequent heat balance calculations, Q_{FZI} will be the value used rather than Q_{FO}.

From Figure 2.13, the stripout from residuum in a 4-tray section is 1.2 vol. percent of the amount fed when the steam rate is 10 lb./bbl. of stripped liquid. Referring to Figure 2.11, it is seen that the total feed to the residuum stripping section is the sum of the feed flash liquid and the overflash.

Overflash = 41.7 bbl./hr.

Feed flash liquid = 537.4 bbl./hr.

Stripping section feed = 579.1 bbl/hr.

Stripout = 0.012 x 579.1 = 6.9 bbl./hr.

Net bottoms = W = 579.1 -6.9 = 572.2 bbl./hr.
= 196,569 lb./hr.

@ 10 lb./bbl., steam to tower = S_W = 5,722 lb./hr.
= 317.9 mol./hr.

Total hydrocarbon vapor leaving flash zone is (F -W +L_O) and is equal to (2083 -572.2 +41.7) = 1,552.5 bbl./hr. which is 74.5 vol. percent and 70.0 wt. percent of the crude feed. Thus, total hydrocarbon vapor from flash zone = 1552.5 bbl./hr.
= 429,994 lb./hr.
@ 47.1° API
= 2,704.4 mol./hr.
@ 159 mol. wt.

Total vapor from flash zone = (hydrocarbon + S_W)
= (2704.4 + 317.9) = 3022.3 mol./hr.

Therefore, the hydrocarbon partial pressure in the flash zone at a total pressure of 25.0 psia. is

P'_{HC} = 25.0 (2704.4/3022.3) = 22.4 psia.

Using Maxwell's nomogram, the 14.7 psia./74.5 vol. percent temperature of 648°F converts to 690°F at 22.4 psia. This is the flash zone temperature.

Assuming a 30°F temperature drop across the stripping section, the bottoms liquid, W, leaves the tower at 660°F. Thus,

Q_W = (196,569 x 427) = 83.935 MMBTU/hr.

Using 65 psia./500°F stripping steam,

Q_{SW} = 5722 x 1283 = 7.341 MMBTU/hr.

In summation, the external heat quantities to the top of the flash zone are,

Q'_{LFZ} = Q_{FZI} + Q_{SW} = 344.349 MMBTU/hr.
Q'_{OFZ} = Q_W = 83.935 MMBTU/hr.
$\Delta Q'_{FZ}$ = $(Q_i - Q_O)_{FZ}$ = 260.414 MMBTU/hr.

3. Crude Evaluation and Product Properties

A. Product Distillations

By a trial-and-error procedure, the whole crude is cut into fractions satisfying the boiling range and

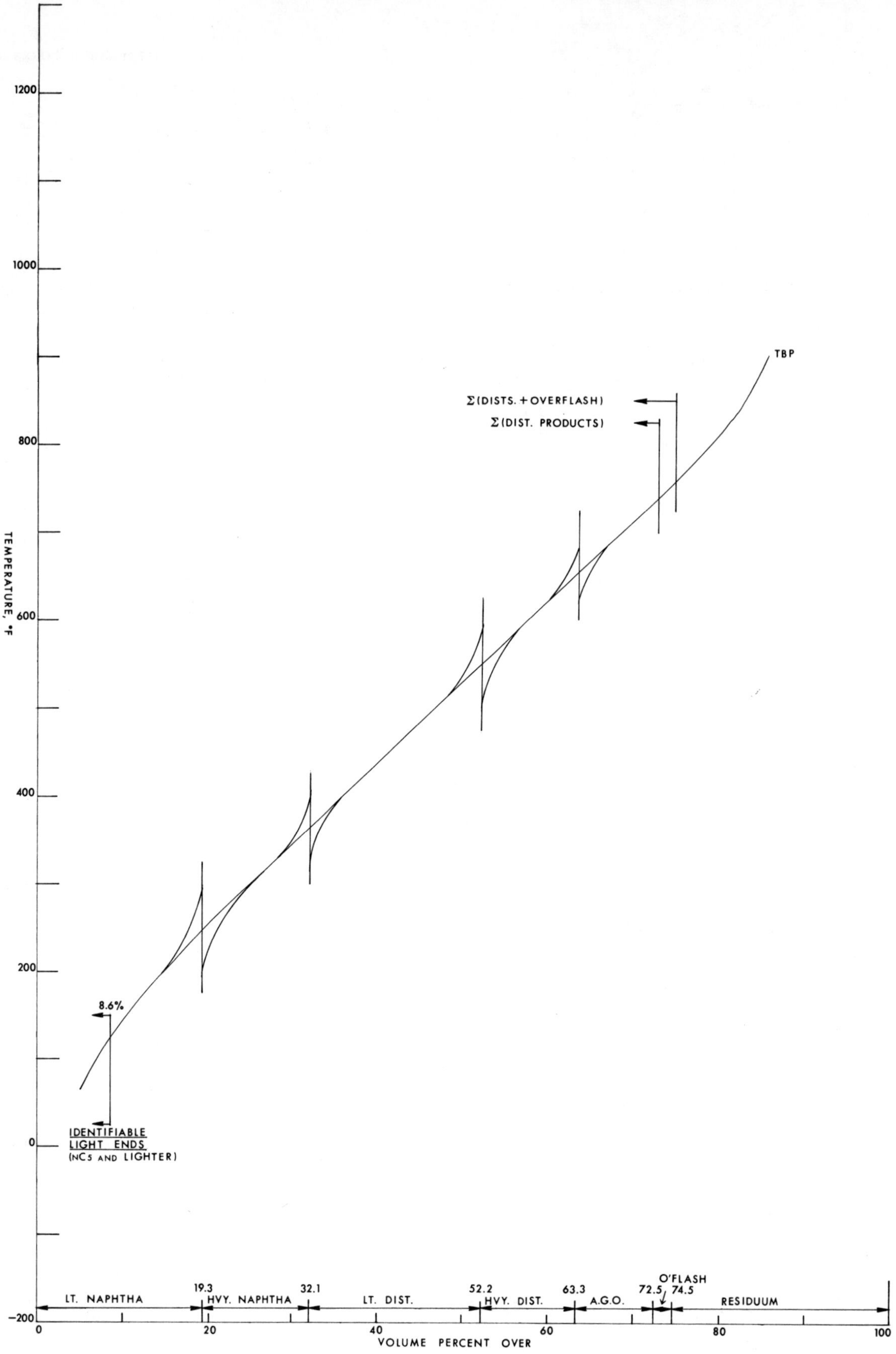

Figure 2.28. 36.3 degree API crude, TBP curves for atmospheric tower feed and products.

(5-95) gap criteria defined in Section 1,B. Figures 2.15 and 2.16 are used to estimate the TBP temperatures of the fractions. Edmister's techniques are used to convert the TBP's to ASTM & EFV distillations. Figure 2.28 shows the proper apportionment and TBP distillations. The data is summarized as follows.

Stream	Vol. %	TBP, °F	ASTM, °F	EFV, °F
Light Naphtha				
	50	140	149	137
	70	187	182	153
	90	237	221	169
	100	295	275	189
Heavy Naphtha				
	0	199	254	278
	10	237	272	287
	30	277	293	297
	50	305	309	304
	70	328	324	310
	90	363	350	321
	100	398	380	330
Light Distillate				
	0	332	386	417
	10	376	408	428
	30	420	432	440
	50	457	454	451
	70	495	480	462
	90	542	517	478
	100	590	560	493
Heavy Distillate				
	0	510	545	594
	10	537	556	598
	30	575	575	608
	50	602	590	615
	70	623	603	621
	90	655	626	630
	100	685	650	637
Atmospheric Gas Oil				
0	0	625	641	698
10	10	647	650	703
30	30	677	655	711
50	50	697	676	716

By plotting the ASTM distillations, the following ASTM (5-95) gap requirements are obtained.

LN - HN: (268 - 242) = +26°F.
HN - LD: (498 - 363) = +35°F.
LD - HD: (550 - 537) = +13°F.
HD - AGO: (645 - 638) = +7°F.

B. Flow Rates and Properties of Product Streams

Stream	bbl./hr.	lb./hr.	mol./hr.	°API	Mol. Wt.
LN*	402.0	94,599	1100.0	78.7	86.0
HN	266.6	72,484	587.7	50.5	123.3
LD	418.7	121,013	654.5	39.7	184.9
HD	231.2	69,413	229.8	33.4	302.1
AGO**	192.3	60,199	122.9	26.6	489.8
W***	572.2	196,569	-	12.6	-

Notes: * - Rates and properties are for the gross overhead product. The properties of the gas and liquid phase into which LN separates in the reflux drum will be determined in Section 3,D which follows.

** - Stripped with steam - 1923 lb./hr.

*** - Stripped with steam - 5722 lb./hr.

HN, LD, and HD are reboil-stripped to a revaporization equivalent to steam stripping of 10 lb./bbl.

C. Flow Rates Around Product Strippers

From Figure 2.13 is estimated the volume percent of feed to product strippers which is revaporized, i.e., the stripout. Summarizing,

Stream	% S.O.	S.O., lb./hr.	F., lb./hr.	Atmos.EFV, °F
AGO	5.0	3168	63,367	695 @ -5%
HD	4.5	3271	72,684	590 @ -4.5%
LD	4.0	5042	126,055	411 @ -4.0%
HN	3.3	2474	74,958	274 @ -3.3%
LN	No	-	-	189 @ +100%

Note that the EFV temperature approximations are obtained by plotting the EFV Data calculated in Step 3,A and extrapolating the front end to minus 20 percent.

D. Analysis of Total Light Naphtha (LN)

By plotting the TBP curve of the total LN going overhead and knowing the light ends composition, one can approximate the amounts present of those components heavier than the heaviest identifiable discrete component. The TBP end point of the light naphtha is 295°F and light naphtha is the front 19.3 vol. percent of the crude.

From the light ends analysis given in Section 1,A,2, we see that C_2H_6 through nC_5H_{12} comprises 8.6 vol. percent of whole crude which is equivalent to 44.6 vol. percent of LN. Considering the LN to be a mixture of paraffinic and naphthenic hydrocarbons, the TBP boiling ranges of C_6 and heavier components can now be defined and, from these can be determined the TBP cut points between components. In tabular form,

Comp.	TBP Range, °F	TBP Cut Point, °F
C_6	122-177	176
C_7	175-218	215
C_8	211-269	262
C_9	255-314	

From this, and using the densities and molecular weights of n-alkanes to define the C_6 through C_9's, we arrive at the following composition of LN.

Component	Vol. Percent	Mole Percent	Moles/hr.
C_2H_6	0.52	0.80	8.8
C_3H_8	4.66	6.61	72.7
iC_4H_{10}	7.26	8.67	95.4
nC_4H_{10}	11.91	14.77	162.5
iC_5H_{12}	10.88	11.61	127.7
nC_5H_{12}	9.33	10.06	110.7
C_6	21.94	20.84	229.1
C_7	17.00	14.40	158.4
C_8	11.50	8.77	96.5
C_9	5.00	3.47	38.2
Total	100.00	100.00	1100.0

The LN is not totally condensible at 15.0 psia and 120°F in the reflux drum. Therefore, the separate properties of the vapor and liquid phases into which LN separates by virtue of equilibrium flash condensation are obtained by making a flash calculation at the conditions of temperature and hydrocarbon partial pressure which exist in the reflux drum.

At 120°F, the vapor pressure of water is 1.692 psia. Thus, since liquid water will be formed, the water will exert its full vapor pressure and thus,

$$P'_{HC} = 15.0 - 1.693 = 13.308 \text{ psia.}$$

At 13.308 psia and 120°F, LN seaprates in the following way

Stream Component	Vapor Phase	Liquid Phase
C_2H_6	8.6	0.2
C_3H_8	68.4	4.3
iC_4H_{10}	83.5	11.9
nC_4H_{10}	135.6	26.9
iC_5H_{12}	87.7	40.0
nC_5H_{12}	69.7	41.0
C_6	85.1	144.0
C_7	28.5	129.9
C_8	6.9	89.6
C_9	1.1	37.1
Total,		
mol./hr.	575.1	524.9
lb./hr.	45,887	48,712
bbl./hr.	195.2	206.8
Mol. Wt.	79.8	92.8
°API	78.9	78.5
Water,		
mol./hr.	73.1	351.2
lb./hr.	1,317	6,328

4. Draw Tray Locations and Estimated Operating Temperatures

Draw tray temperatures are estimated using Figure 2-18 for the EFV temperatures cited in Section 3,C.

Stream	Draw Tray No. (from bottom)	Temperature, °F
AGO-D1	8	610
HD-D2	14	560
LD-D3	20	425
HN-D4	28	295
LN-(D5 + V5)	35	205

Note that linear temperature and pressure variation between draw trays is assumed so that, for the purpose of simplifying this illustration, we will not plot a temperature-pressure profile.

5. Calculation of the First Draw Tray Above the Flash Zone (AGO Tray - No. 8)

Refer to Figures 2.17 and 2.19 which detail the calculational procedure to be followed.

Leaving Tray 8 at an assumed temperature of 610° are

357,509 lb./hr. total hydrocarbon vapor @ 51.0° API.
60,199 lb./hr. hydrocarbon liquid product @ 26.6° API.
5,722 lb./hr. steam.

Thus,

$$Q_{V8} = (357,509 \times 538) + (5722 \times 1342).$$
$$Q_{L8} = (60,199 \times 417).$$
$$Q_{R8} = \Delta Q'_{FZ} - (Q_{V8} + Q_{L8}) = 260.414 - 225.122 = 35.292 \text{ MMBTU/hr.}$$

The liquid which is to be revaporized in the sidestream stripper falls from Tray 9 at 603°F and, in passing across Tray 8 is heated to 610°F, thus absorbing 0.016 MMBTU/hr.

The heat removal capability of Tray 9 reflux from 603°F liquid to 610°F vapor at 26.6° API is 94 BTU/lb. Thus,

$$L_9 = (Q_{R8} - 0.016)/94 = 375,277 \text{ lb./hr.} = 766.2 \text{ mol/hr.}$$

In the vapor leaving Tray 8, we neglect the pressure of HD (D2) and consider all other lighter products (D3 - D5) plus the steam from the bottoms stripper to be inerts. Thus, the hydrocarbon partial pressure of D1 in the vapor leaving Tray 8 is calculated as

$$P'_{HC} = 24.35 \, (766.2/3426.3) = 5.44 \text{ psia.}$$

Using Maxwell's nomogram, the estimated 14.7 psia bubble point (695°F) of the unstripped liquid on the tray, when converted to 5.44 psia is 610°F which checks the assumed temperature.

At the product stripper, D1 will be stripped with 1923 lb./hr. of 65 psia/500°F steam and will leave the tower 30°F cooler at 580°F. Thus,

$$Q_{S1} = 1923 \times 1283 = 2.467 \text{ MMBTU/hr.}$$
$$Q_{D1} = 60,199 \times 393 = 23.658 \text{ MMBTU/hr.}$$

Now, the heat quantities to the base of Tray 9 are

$$Q'_{i8} = Q_{iFZ} + Q_{S1} = 346.816 \text{ MMBTU/hr.}$$
$$Q'_{o8} = Q_{oFZ} + Q_{D1} = 107.493 \text{ MMBTU/hr.}$$
$$\Delta Q'_8 = 239.323 \text{ MMBTU/hr.}$$

These values are tabulated on the Heat and Material Balance Worksheet Summary

By similar techniques which are detailed in the text, the temperatures of Trays 14, 20, and 28 are calculated as 559, 424, and 293°F. Since these streams are reboiled, they leave their respective towers 30°F hotter than their draw tray temperatures.

The heat effects up through Tray 28 are summarized on the Worksheet.

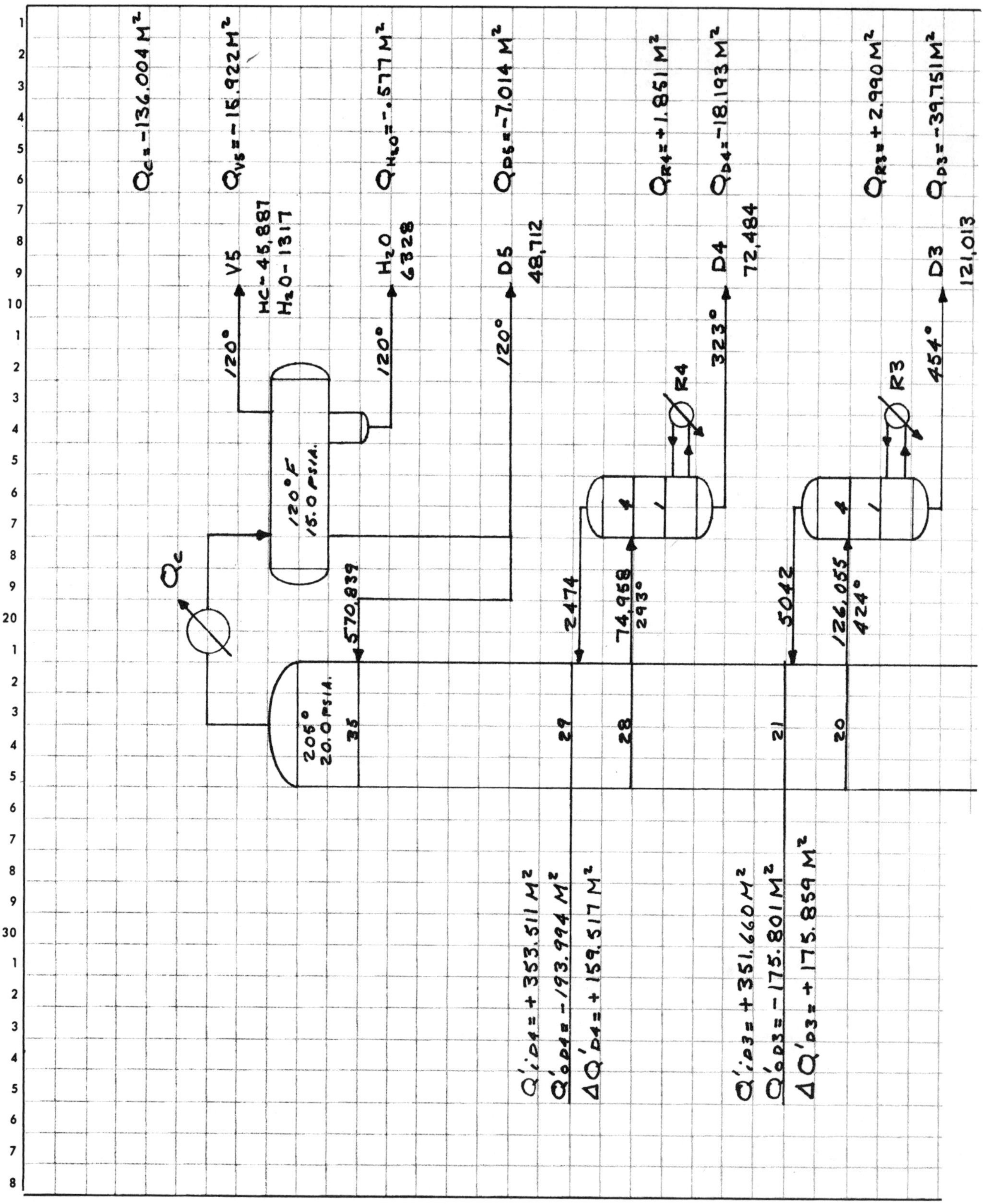
Q_C = -136.004 M²
Q_V5 = -15.922 M²
Q_H2O = -.577 M²
Q_D5 = -7.014 M²
Q_R4 = +1.851 M²
Q_D4 = -18.193 M²
Q_R3 = +2.990 M²
Q_D3 = -39.751 M²
V5
HC - 45,887
H2O - 1317
H2O
6328
D5
48,712
D4
72,484
D3
121,013
120°
120°
120°
323°
454°
R4
R3
120°F
15.0 PSIA.
Q_C
570,839
2474
74,958
293°
5042
126,055
424°
205°
20.0 PSIA.
35
29
28
21
20
Q'_iD4 = +353.511 M²
Q'_oD4 = -193.994 M²
ΔQ'_D4 = +159.517 M²
Q'_iD3 = +351.660 M²
Q'_oD3 = -175.801 M²
ΔQ'_D3 = +175.859 M²

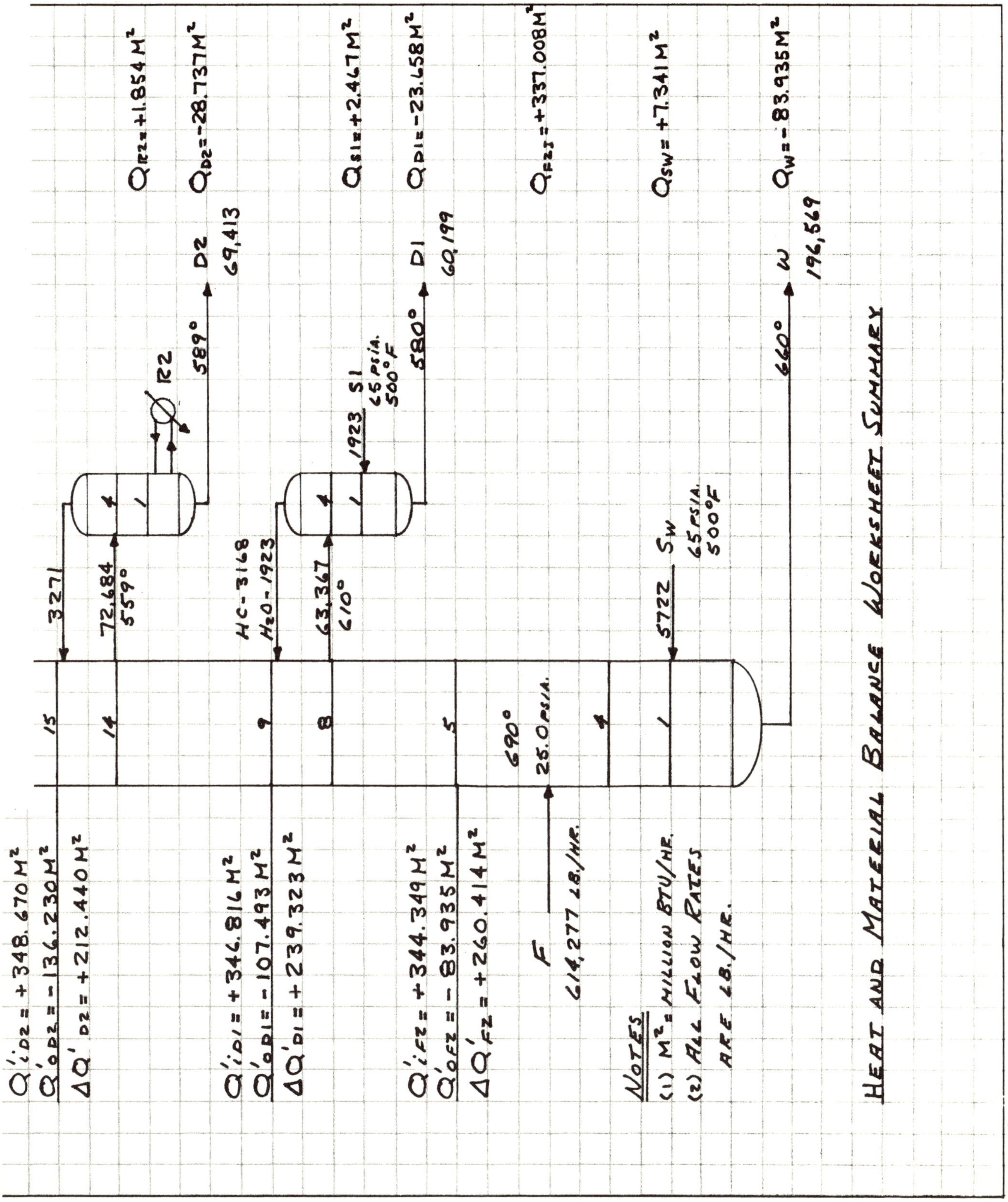
HEAT AND MATERIAL BALANCE WORKSHEET SUMMARY
Q'iD2 = +348.670 M²
Q'oD2 = -136.230 M²
ΔQ'D2 = +212.440 M²
Q'iD1 = +346.816 M²
Q'oD1 = -107.493 M²
ΔQ'D1 = +239.323 M²
Q'iFZ = +344.349 M²
Q'oFZ = -83.935 M²
ΔQ'FZ = +260.414 M²
F
614,277 LB./HR.
690°
25.0 PSIA.
NOTES
(1) M² = MILLION BTU/HR.
(2) ALL FLOW RATES ARE LB./HR.
15
14
9
8
5
4
1
3271
72,684
559°
HC-3168
H2O-1923
63,367
610°
5722
SW
65 PSIA.
500°F
1923
S1
65 PSIA.
500°F
R2
589°
D2
69,413
580°
D1
60,199
660°
W
196,569
QR2 = +1.854 M²
QD2 = -28.737 M²
QS1 = +2.467 M²
QD1 = -23.658 M²
QFZ = +337.008 M²
QSW = +7.341 M²
QW = -83.935 M²

6. Top Tray and Condenser Calculations

Assuming Tray 235 at 205°F, the heat leaving Tray 35 is

Q_{V35} = (94,599x 350) + (7645 x 1153) = 41.924 MMBTU/hr.
Q_{R35} = $\Delta Q'_{28}$ - Q_{V35} = 117.953 MMBTU/hr.
$(q'_R)_{LR}$ = 206 BTU/lb.
L_R = 570,839 lb./hr. = 6151.3 mol./hr.

Neglecting V_5,

P'_{HC} = 20.0 (6676.2/7100.5) = 18.80 psia.

Converting the atmospheric EFV dew point of the total LN to 18.80 psia by Maxwell's nomogram, the assumed temperature of 205°F checks.

Thus, the consider duty is now calculated as

Q_C = (665,438 x 350 + 7645 x 1153)
-(45,887 x 315 + 1317 x 1114 + 6328 x 88 + 48,712 x 144 + 570,839 x 144)
= 136.024 MMBTU/hr.

By overall system heat balance,

Q_C = Q_{28} -(Q_{V5} + Q_{D5} + Q_{H_2O}) = 136.004 MMBTU/hr.

ΔQ_C = 0.020 MMBTU/hr. = .0147% deviation.

7. Fractionation

By making heat balances to the bases of draw trays and the top tray, the following fraction criteria can be checked.

Separation	LD/P	No. Trays	F	(Δt_{50}) ASTM	(5-95) Possible
LN-HN	8.15	8	65.2	160	+46
HN-LD	4.33	8	34.7	231	+45
LD-HD	2.18	6	13.1	278	+36
HD-AGO	1.16	6	7.0	314	+26

All the above exceed the fractionation requirements.

vacuum tower

The objective of crude oil distillation is generally to maximize the extraction of distillate liquids from the raw crude feed. Since distillates of given gravity are cleaner and freer from objectionable impurities than an unprocessed crude having the same gravity, they command a higher price in the marketplace. Additionally, these distillates may serve as feedstocks for other process units where they can be converted to even more valuable materials. Thus, it is clearly to the advantage of the refiner to recover as much distillate as possible from a given crude. There are exceptions to this as in the case of asphalt manufacture or where large volumes of residual fuels are required such as for bunkering operations. On the whole, however, this is a fair statement covering most situations.

In an atmospheric tower operating at as low a pressure and as high a temperature as practical in the flash zone, the maximum amount of oils which will vaporize is described approximately by whole crude TBP cut points between distillates and residual liquid of 700 to 800 degrees F. At these operating conditions, the atmospheric residuum, commonly called reduced crude, still contains a large volume of distillable oils which can be recovered by vacuum processing. At the current economic operating levels of maximum vacuum and temperature in the flash zone, whole crude TBP cut points between distillates and residuum can be as high as 1,075 to 1,125 degrees F. This cut point is strongly influenced by the metals content of the distillates, particularly nickel and vanadium. For a 30 degree API crude, the vacuum distillates can be as much as 30 volume percent of the whole crude.

Types of Operations in Vacuum Distillation

Vacuum distillates and residua can have several uses, depending upon the type of crude oil feed, the type of refinery and its downstream processing capabilities and, most important of all, the anticipated sales of products. Thus, the general function of the vacuum tower is to remove the maximum possible amount of distillate from the charge stock consistent with meeting product specifications on the residuum as well as the distillates. The vacuum gas oils have several uses, and, although they all will have the same general properties, each will require different considerations in design. Some of these problems are discussed briefly as follows.

The principal criteria for producing lube oil fractions are viscosity, the rejection to the residuum of heavy impurities, color and, to a certain extent, boiling range, but not necessarily to the same degree of importance as in atmospheric distillates. These stocks must be as free as possible of Conradson carbon, but metals content, while an indication of poor fractionation, is not a stringent requirement in itself. These oils are usually refined further by solvent extraction and other types of treating including hydroprocessing. In the course of these processing steps, they are often redistilled.

Gas oils intended for catalytic cracking feedstocks must be very clean and require a good separation between distillate and residuum. These oils should have the minimum amount of Conradson carbon content without sacrificing gas oil recovery. This is necessary to prevent excessive laydown of coke on the cracking catalyst. The metals content, particularly vanadium and nickel, must be strictly limited since these are severe catalyst poisons.

Feedstocks for distillate hydrotreaters and/or hydrocrackers need not necessarily be metal free since these same metals are often found in hydrogenation catalysts. It is imperative, however, that they be as free as possible of carbon and asphaltic materials. If they are not, coking of the catalyst will occur more rapidly, and more frequent regeneration will be required. Product cleanliness is far more important here than in the case of catalytic cracking feedstocks.

Distillate fuel oils which are to be used without further processing are usually specified by API gravity, viscosity, metals content and flash or fire point. Since the latter two properties are general functions of the front end boiling range and the others of total stream yield, atmospheric boiling range, either TBP or ASTM, can be used to define the separation. Metals content, particularly vanadium, is a source of ash and is also harmful to most furnace refractory materials.

Vacuum tower residuum properties can be set in various ways. When distillate production is to be maximized, the amount of gas oil allowed to remain in the bottoms stream must be minimized, and this material, usually 0 to 5 degrees API, is blended into residual fuels. This is called *pitch operation.* In *asphalt operation,* some gas oil must be left in the residuum in order to provide the proper degree of plasticity. The gravity of asphalt streams usually falls in the range of 5 to 8 degrees API. Not all crudes can be used to make asphalt since this is a question of whether or not the particular residuum has the proper composition of matter to render it suitable in asphalt applications. This question can only be answered experimentally and is usually contained in a detailed crude assay. Pitch operation can be practiced with any type of crude.

For pitch operation, one can normally set the volume percent of either the whole crude or reduced crude which is to be yielded as vacuum residuum. This is, in effect, specifying a TBP cut point for vacuum distillates. In some cases, the vacuum unit is defined in exactly this way. It can also be defined by determining the amount of distillate yield which will produce a bottoms stream having the desired gravity. More usually, though, the limiting factor is the practical economic limit to how much distillate can be lifted from a given system and is defined in terms of the lowest possible operating pressure and highest allowable oil temperature in the tower flash zone.

For asphalt operation, experimental data is necessary, relating asphalt penetration to residual volume. Using this information, distillate volume can be readily established. Units have been built to make asphalt over the penetration range of 85 down to 10 (approximately 7 to 4 degrees API). The asphalt specifications will most often be set by the owner, and it may be necessary to design for more than one grade of product.

One other method of vacuum tower operation which is rather rare is the production of large volumes of bunker fuels. This operation is usually practiced in foreign refineries in the producing areas where marine fuel requirements are high. These refineries usually produce distillates to satisfy marketing requirements which may vary a great deal and yield the remaining gas oil with the residuum. The resulting residual fuels are slightly lighter than in normal operations but are still quite adequate from the end use standpoint.

In meeting these various product specifications, it must be remembered that the key to satisfactory operations is the maximum volume of clean gas oil, free from contamination by heavier materials. In pitch operation, the quality of the residuum is of little importance other than to minimize its content of recoverable oils. In asphalt operation, gas oil quality must remain high, but the residuum must meet certain stringent specifications as well.

Lube or Specialty Vacuum Distillate Operations

Vacuum towers for manufacture of oils destined either for lube oils or some other type of specialty oils must be designed to provide the same relative degree of fractionation between streams as in the atmospheric tower. Accordingly, the two types of towers are designed and calculated in approximately the same way. Sidestream products are stripped in external towers to control front end properties. Heat recovery is practiced by exchanging hot downflowing tower liquid streams against incoming cool crude oil feed for the atmospheric tower or against other cooler process streams in the unit. This heat recovery can be practiced in two different ways.

Figure 3.1 shows utilization of cooled pumpback reflux from the draw tray to the tray below at all trays except the top sidestream. This liquid is condensed by pumparound heat removal using a grid type material for vapor-liquid contacting. Cooled reflux is pumped back to the tray below the draw tray to provide fractionation between the two light vacuum distillates.

Figure 3.2 shows utilization of cooled pumparound reflux in two sections of the tower. In terms of heat removal and equipment requirements, the two methods are relatively equivalent. Both the external heat recovery and the internal vapor and liquid traffic quantities are essentially equal. The pumpback reflux system has the advantage

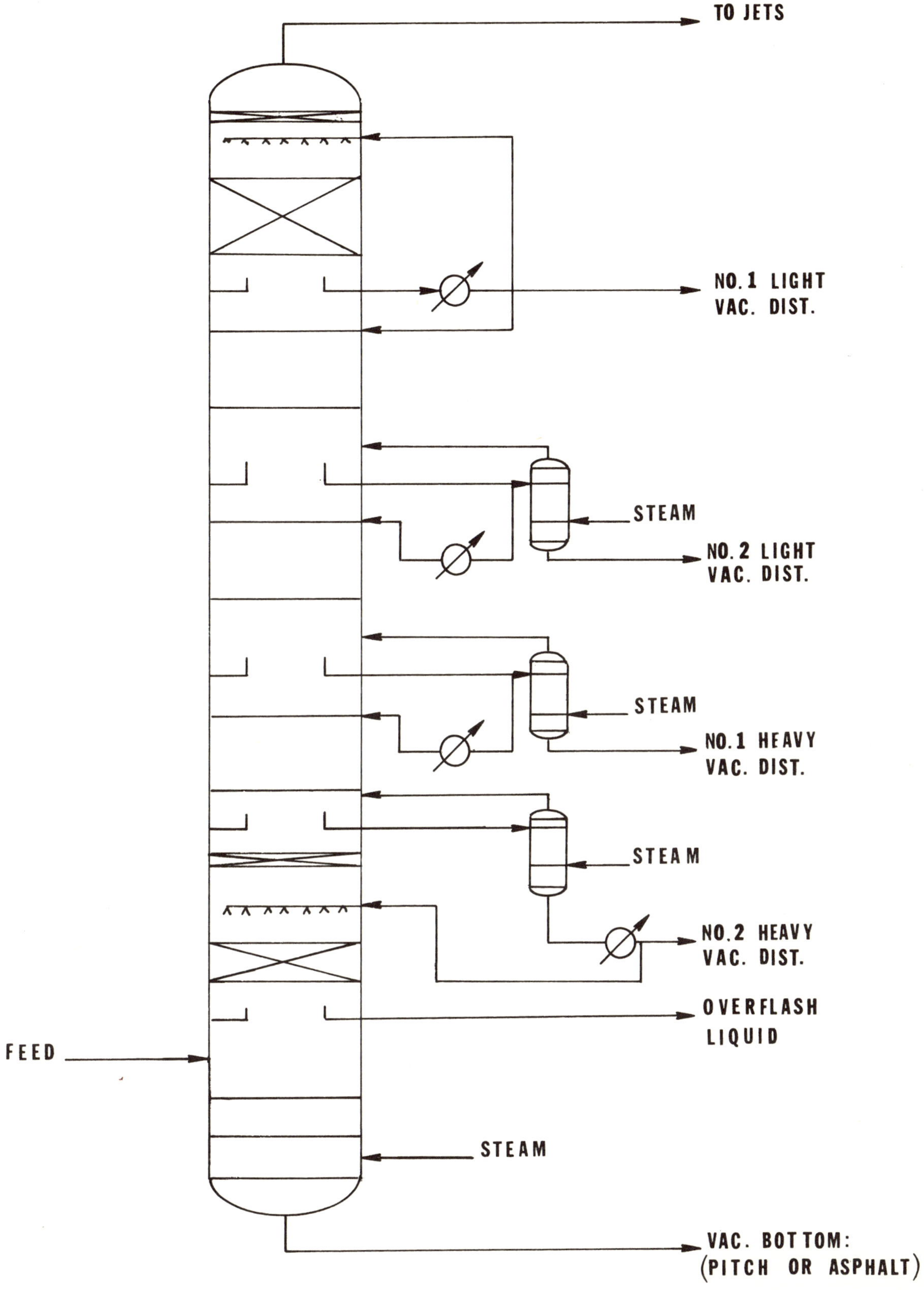

Figure 3.1. Lube type vacuum tower with pumpback reflux heat removal.

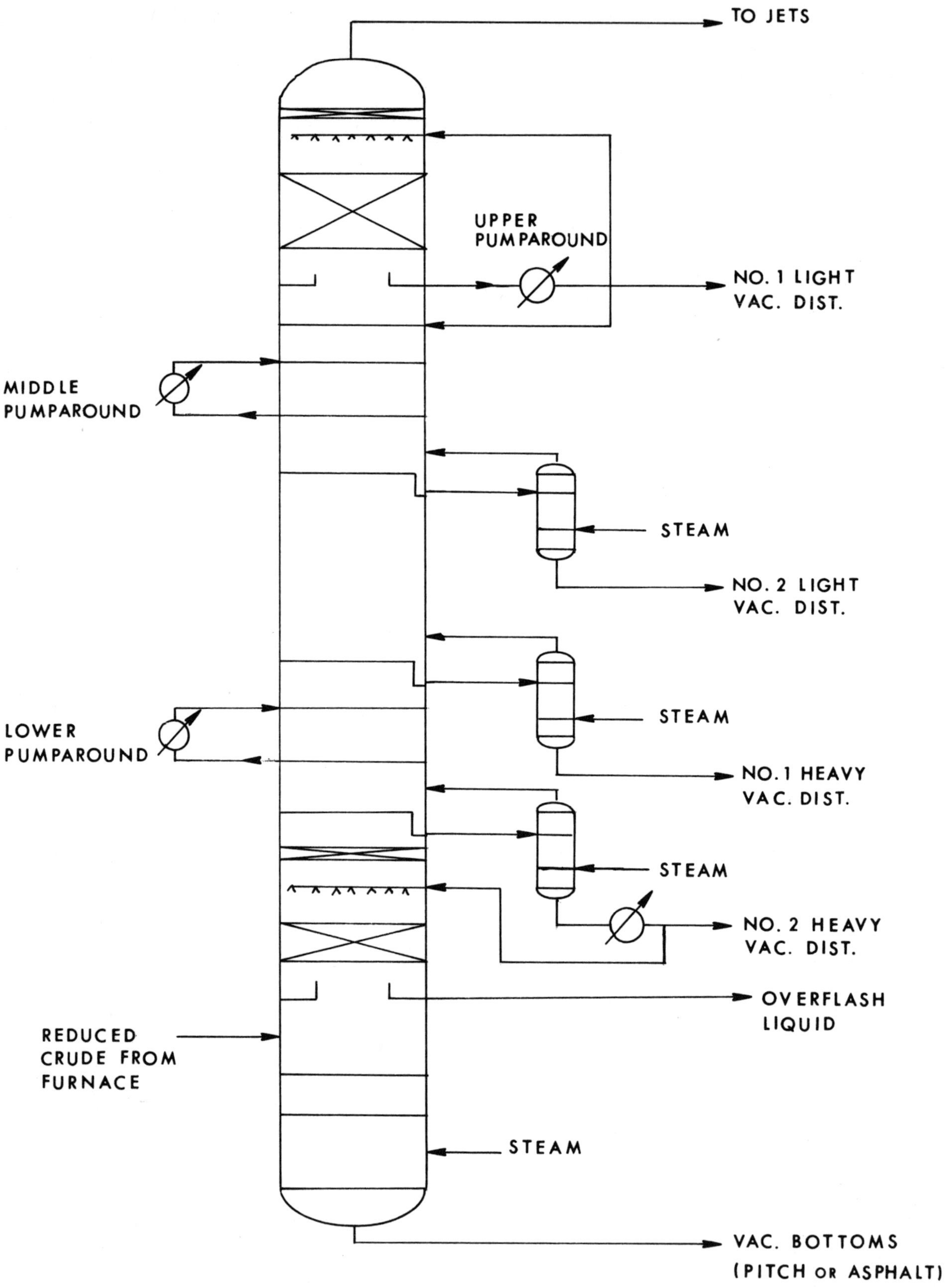

Figure 3.2. Lube type vacuum tower with pumparound reflux heat removal.

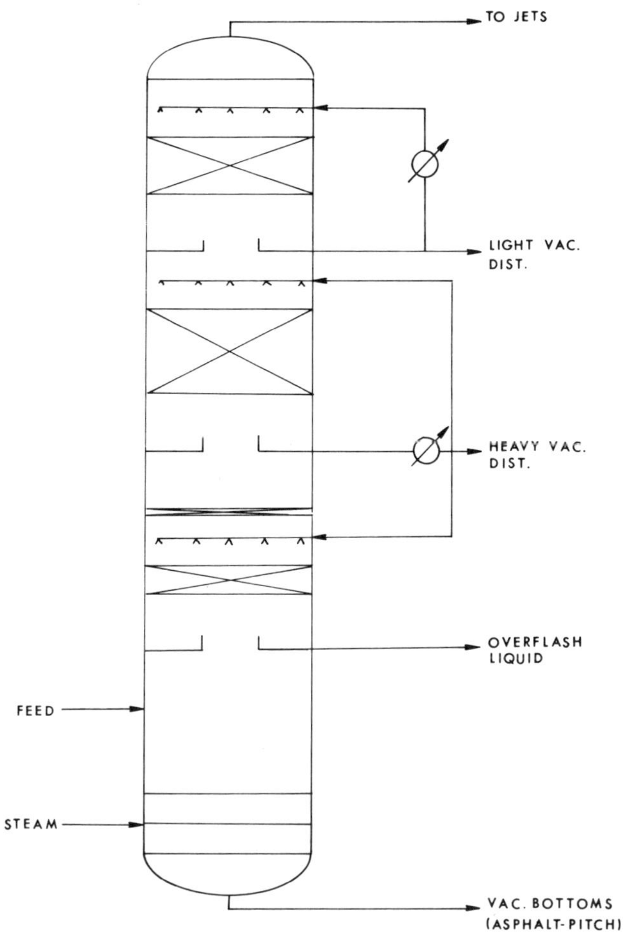

Figure 3.3. Fuels type vacuum tower.

that all trays are true fractionating trays since all internal liquids provide equilibrium reflux whereas the liquids in the pumparound sections are foreign with respect to the composition of the vapors passing through these sections. For this reason, a pumparound section tray is usually considered to be only 30 to 50 percent as effective as other trays in the tower. Thus, a tower with pumparound reflux must either be provided with more trays, usually one in each section, or sacrifice a certain amount of fractionation capability. Since trays require pressure drop to be effective, anything which minimizes tray requirements is directionally correct.

Note the two different types of draw trays shown in the illustrations. Figure 3.1 uses the chimney type in which the vapor flows through risers, and the liquid is collected on the tray deck. Figure 3.2 utilizes a fractionating tray with a sealed downcomer in which the liquid is collected. Obviously, a Figure 3.2 arrangement will provide more trays for a given pressure drop although a Figure 3.1 arrangement is a better way of collecting and removing the liquid. Detailed design considerations for chimney trays have been presented in a recent article (1). Tray design principles for low pressure drop applications are adequately discussed in various manufacturer's bulletins (2,3). In recent years, the trend has been toward the greater use of sieve trays in vacuum towers because of their lower cost and inherently lower ultimate minimum pressure drop. Bubble-caps are rarely used any longer.

The number of trays between draw trays is set rather arbitrarily. Nelson (4) recommends using 3 to 5 trays between the draws. If one employs fractionating type draw trays, this would indicate four trays between draws at the maximum since the draw counts as a tray. If chimney type draw trays are used, one more tray in the section would be required to provide the equal degree of fractionation.

Fuels Operation

The manufacture of distillates either directly for fuels or for feed to downstream processing units ordinarily does not require any particular degree of fractionation between cuts. Also, wide cuts are usually acceptable. For these reasons, the distillates can be condensed by cooled pumparound reflux, grid type contacting sections and chimney draw trays. For all practical purposes, the operation of the main condensing sections can be described as a single-stage equilibrium condensation.

Figure 3.3 shows a typical fuels type tower. Note that the overflash liquid is condensed by cooled pumpback reflux rather than by another pumparound circuit. There are two reasons for this. First, another pumparound circuit would require significant additional investment. Secondly, it is generally believed that the materials which contaminate the heavy vacuum gas oil are vaporized upward from the flash zone rather than being entrained upward. Accordingly, product quality is more easily maintained by refluxing this section rather than by a high density washing with cool oil.

The grid materials used in these towers have been developed only recently. The process design of the grid sections is discussed thoroughly in the manufacturers' literature (5).

Economic Considerations in Vacuum Tower Design

In the design of any vacuum tower, the first question to be settled is the selection of the optimum operating pressure of the system. In order to simplify this discussion, let us consider certain facts, assuming that a maximum allowable flash zone temperature has been set.

1. Lowering the hydrocarbon partial pressure in the flash zone increases vaporization and, hence, distillate production.
2. Lowering the total pressure of the system decreases the amount of steam required to effect a given vaporiza-

tion. At a low enough pressure, it is theoretically possible that no steam would be required. Note that the steam to a vacuum tower is for the primary purpose of reducing the hydrocarbon partial pressure in the flash zone and not for stripping the vacuum bottoms. The base sections of vacuum towers, however, are normally provided with stripping trays.

3. Raising the system pressure increases steam requirements and also increases tower cross-sectional area requirements.
4. Lowering the system pressure increases the steam requirements for the vacuum producing jets.

From this, it follows that the optimum pressure will be that which minimizes total steam usage. This involves a study of requirements for vapor pressure reduction steam and jet steam at various pressure levels. Vacuum jet vendors can be very helpful in this type of study. If such assistance is not available, an excellent presentation of this subject has been made by Ludwig (6).

At present day economic conditions, the minimum practical pressure at the top of the tower is in the range of 10 millimeters Hg absolute.

Vacuum Unit Charge Data

Since the reduced crude charge to a vacuum tower has already been partially processed in the atmospheric tower, some basic information about the whole crude and its fractions has been developed and should be available at this point. However, it is necessary to develop feed flash data from the reduced crude apart from that previously calculated during the design of the atmospheric section.

It is highly desirable to base the development of vacuum region equilibrium flash data upon experimental vapor-liquid equilibrium studies. Since the design of multimillion-dollar facilities hinges on the accuracy of the basic data, it seems recklessly penurious to skimp on the relatively small costs involved in acquiring sound basic design information. However, some still believe that a whole crude TBP curve and gravity study is sufficient for the entire design. This may be a defensible position for feasibility studies or for rough order-of-magnitude designs since a properly executed design will include all the required process systems. There well may be errors in material and heat balances and in equipment sizing, but these tend to be rather small when compared to the error inherent in failing to provide all the necessary process components. However, for construction grade design work, the engineer should insist on having available experimental vacuum region vapor-liquid equilibrium data.

In the following discussion, the analytical procedures for converting the atmospheric TBP curve of the reduced crude into a vacuum region phase diagram are described. Although most of the physical property data for the reduced crude and its product fractions can be obtained from the assay of the whole crude, the determination of the reduced crude TBP curve and its gravity must still be measured in the laboratory. Given this basic data, the following three-step procedure is used to develop all the necessary information.

1. Using Packie's (7) procedure, calculate the atmospheric EFV curve of the reduced crude.
2. Using Edmister's (8) procedures, convert the atmospheric EFV curve to various levels of reduced pressure down to 5 millimeters Hg absolute. Plot a vacuum region phase diagram.
3. Using data from Maxwell (9) and/or Winn (10), calculate the average molecular weight and the characterization factor, K, for the whole reduced crude.

The physical properties of the various products to be yielded from the tower are calculated or estimated from the properties of the whole crude as given in the assay report. If, for some reason, a full detailed assay is not available, the necessary properties can be estimated by assuming that the value of K is constant for all distillates.

Estimate of Material Balance

The general criteria for establishing the design material balance were outlined in the opening section. The exact material balance for design purposes is determined by calculation of either how much oil can be lifted from the charge at the optimum flash zone temperature-pressure or the relative distillate-residuum separation required to produce a desired asphalt quality. The relative yields of the various distillate fractions will almost invariably be determined by the owner's specifications. For this reason, no discussion of the properties of vacuum distillates is offered here.

Two illustrations are presented with this section, and these present the general procedures for setting up the overall material balance. The first covers a lube-asphalt operation and the second, a fuels-pitch operation. The techniques for each case are quite different and are discussed separately for this reason.

Lube-Asphalt Operation

The final material balance superimposed on the whole crude TBP curve is shown by Figure 3.4. This is for a typical operation producing as basic products asphalt and two lube stocks. The other three distillates are yielded only as determined by the lube cuts. The following procedures were used to develop the overall material balance.

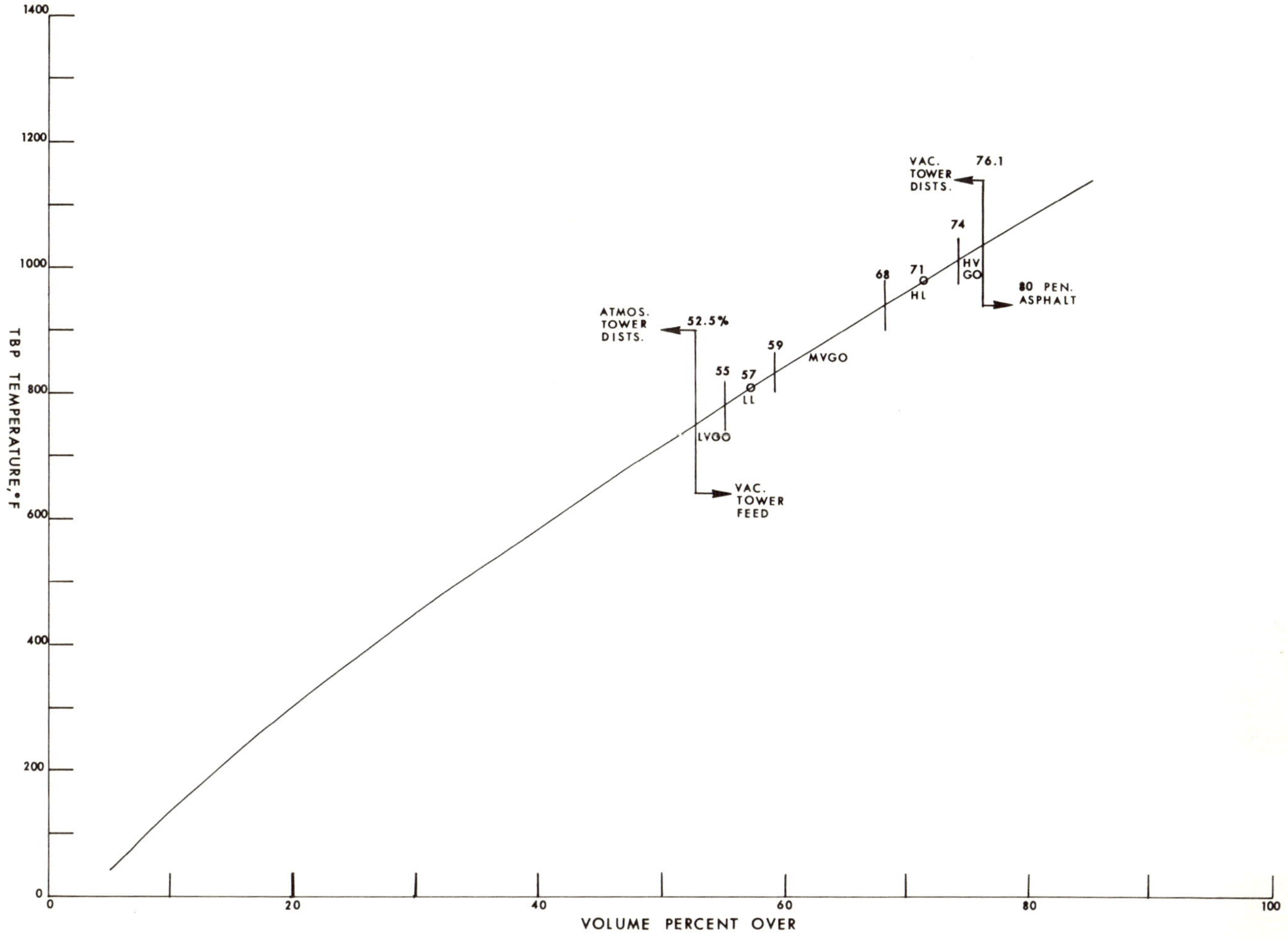

Figure 3.4. Whole crude TBP temperature–yield curve, lube–asphalt operation.

The owner specified the requirement to make three specification products off the tower.

1. Asphalt having a specification of 80 pen. as measured at 77 degrees F.
2. A heavy lube cut having a mid-volume viscosity of 700 SSU as measured at 100 degrees F and a yield on whole crude of 6.0 volume percent.
3. A light lube cut having a mid-volume viscosity of 100 SSU as measured at 100 degrees F and a yield on whole crude of 4.0 volume percent.
4. The required distillation ranges for the lube cuts were specified.

The overall system material balance is established by the following sequence.

1. Total distillate yield is set by analysis of the portion of the crude assay covering asphalt properties. From Figure 3.5, it is seen that an 80 pen. asphalt requires a residuum yield of 23.9 volume percent on whole crude. Thus, the whole crude cut volume between vacuum distillates and residuum occurs at 76.1 volume percent.
2. From Figure 3.6, it is determined that the mid-volumes of heavy lube and light lube, respectively, are 71.0 and 57.0 volume percent of whole crude. From these points and their respective yield requirements, it is determined that the heavy lube will occupy the whole crude range of 71.0 ± 3 volume percent and that the light lube will occupy the whole crude volume range of 57.0 ± 2 percent.
3. From Figure 3.4, it is seen that the TBP cut volume between atmospheric distillates and reduced crude is 52.5 percent. Thus, the tower will also yield the following three product streams.
   a. A light vacuum gas oil (LVGO) over the range 52.5 to 55.0 volume percent.

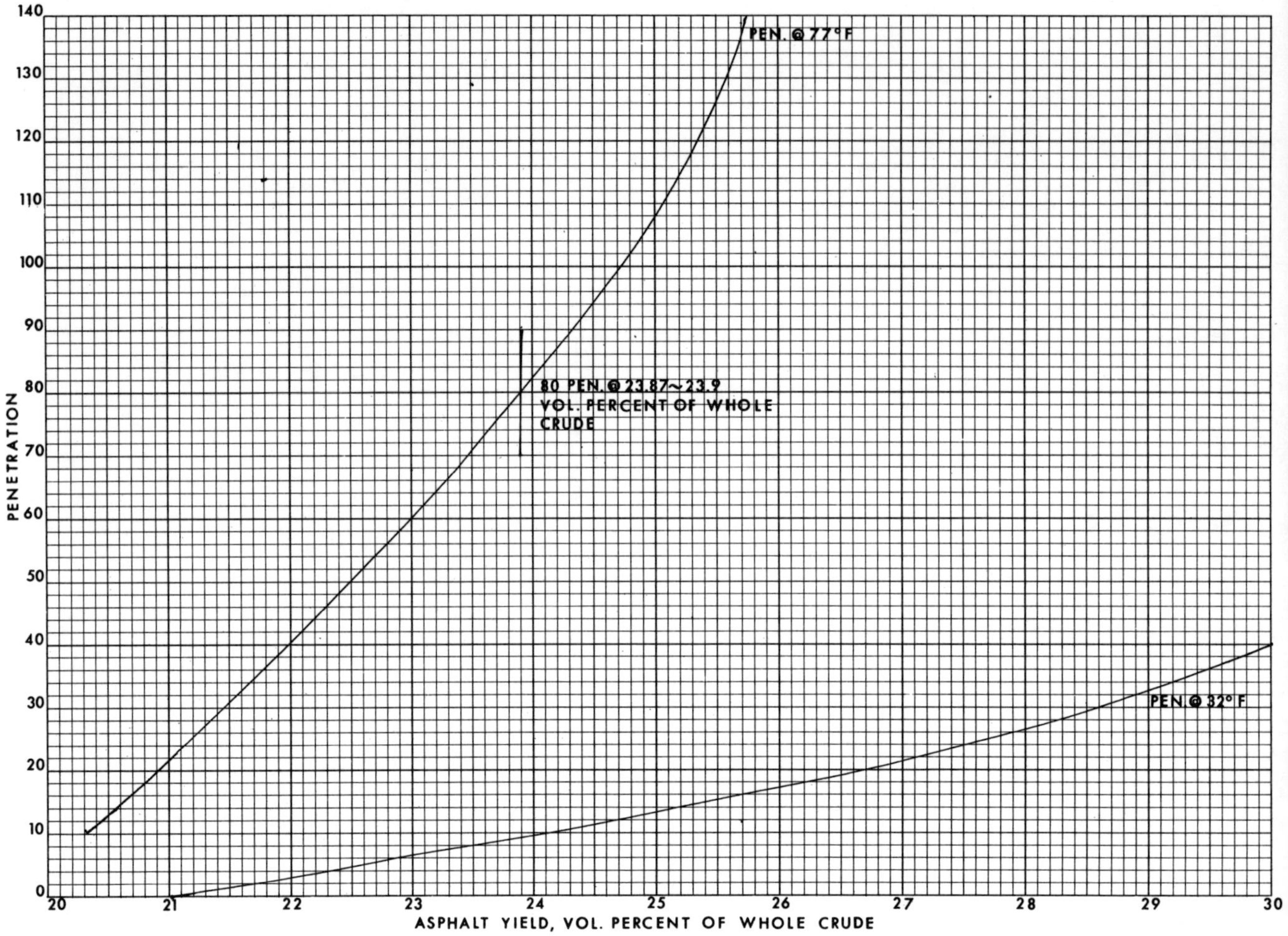

Figure 3.5. Crude assay data. Asphalt yield versus penetration.

b. A medium vacuum gas oil (MVGO) over the range 59.0 to 68.0 volume percent.

c. A heavy vacuum gas oil (HVGO) over the range 74.0 to 76.0 volume percent. This cut may also be thought of as condensed overflash, even, though it is somewhat greater in volume than would normally be designed as overflash.

Thus, the yields of all products have been set. Product properties are determined as follows.

4. From the distillation ranges of the lube cuts, fractionation requirements are established by assuming that the TBP cut point is the mid-point of the total TBP overlap between products.
5. From the TBP distillations of the products, ASTM and atmospheric EFV curves are calculated and plotted. Note that the EFV curves are extrapolated to minus 20 percent for estimating the front end of the unstripped liquids which exist on the draw trays.
6. Stream gravities are obtained from the whole crude properties, and molecular weights are calculated.

Fuels-Pitch Operation

The final material balance superimposed on the whole crude TBP curve is shown by Figure 3.7. This is for a typical operation producing maximum distillates and minimum residuum from a reduced crude feed The following procedures were used to develop the overall material balance.

The owner specified two requirements for the design of the system.

1. The TBP cut point between the distillates and the residuum was set at 1,100 degrees F.
2. The light vacuum gas oil (LVGO) was set at 30 volume percent of the total vacuum gas oils.

The overall system material balance is established by the following sequence.

1. Item 1 above sets the total distillate yield at 81.2 volume percent of whole crude.

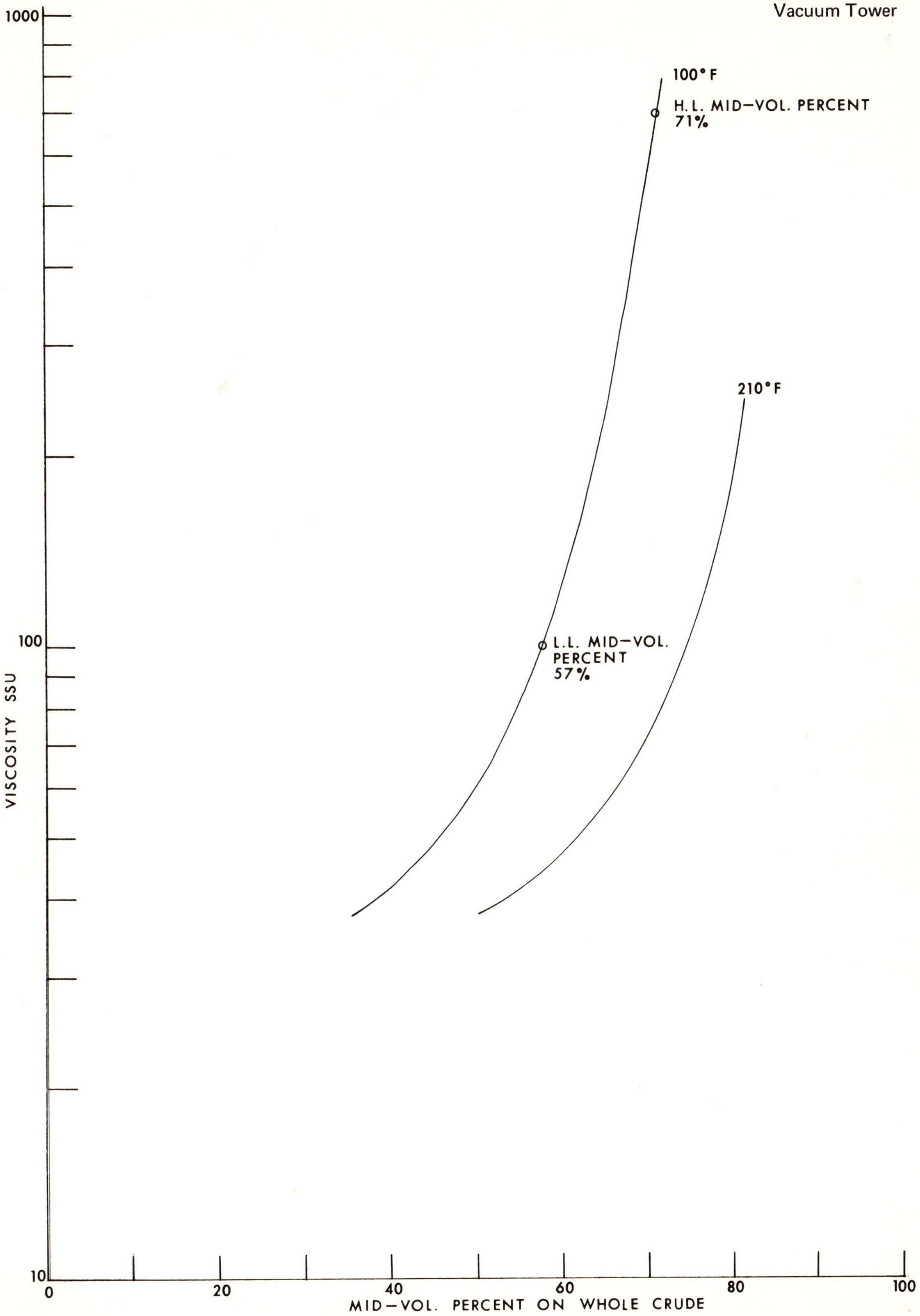

Figure 3.6. Crude assay data, viscosities of vacuum gas oils.

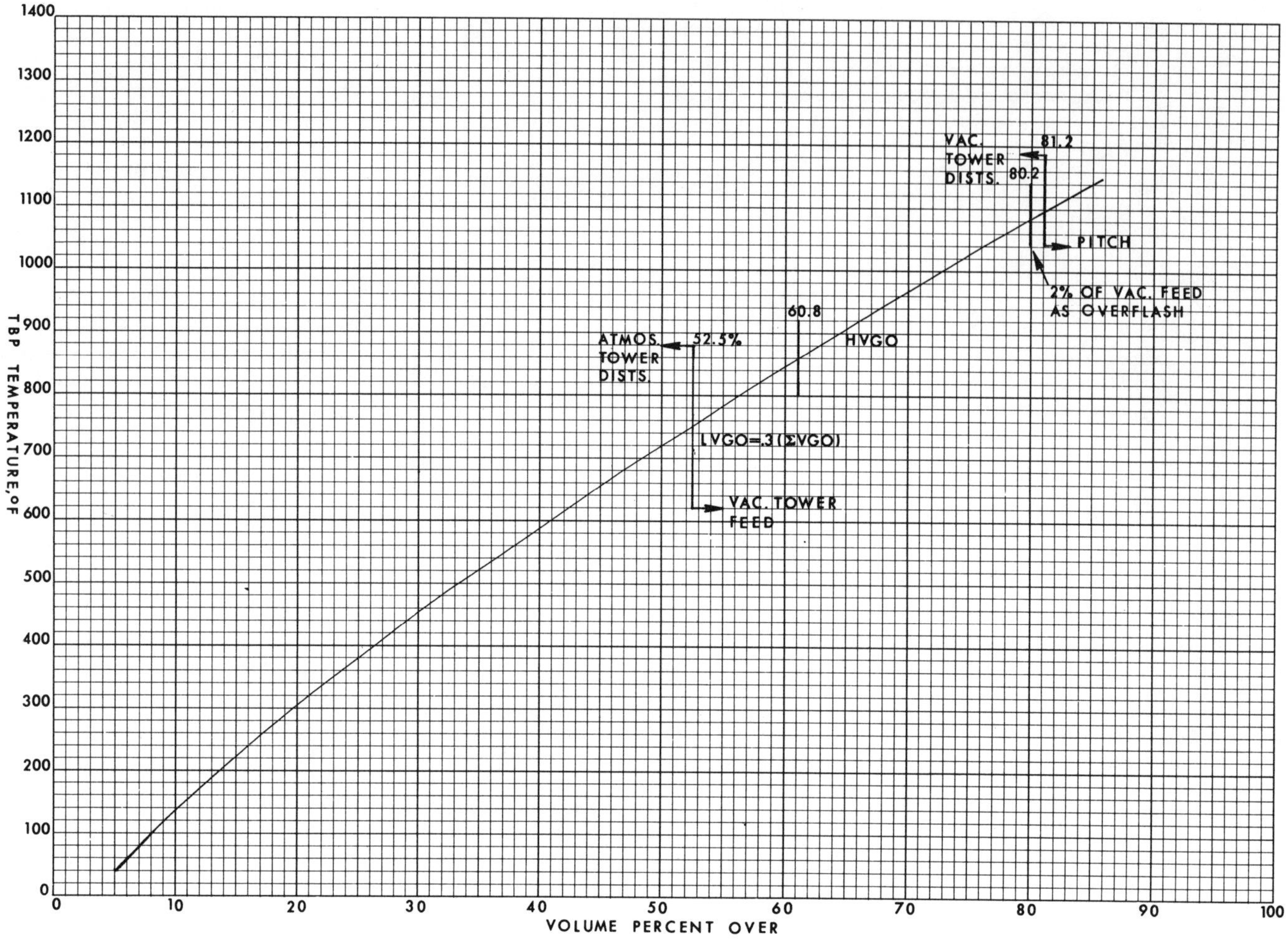

Figure 3.7. Whole crude TBP temperature—yield curve, fuels—pitch operation.

2. The designer sets the amount of overflash depending upon the degree of purity required in the heavy vacuum gas oil. If color requirements and/or allowable metals level specifications are not severe, 1 to 2 volume percent of vacuum tower feed is normally taken as overflash. For tight specifications or for crudes having high metals content, as much as 4 percent may be taken. Two percent was used in Figure 3.7. In practice, operators never take any more overflash than necessary; thus, overflash capability may be considered as built-in extra capacity. Facilities should always be provided for measuring overflash and for yielding it as a product stream, even though it may be returned to the column sump or recycled to the furnace.
3. Item 2 above sets the sidestream yields of heavy and light vacuum gas oils.
4. Since a fuels-type tower operates more or less as a series of equilibrium condensations, the properties of the products are estimated by the following technique.
   a. From the atmospheric EFV curve of the reduced crude, tabulate the EFV curves of the products and convert them to the ASTM and TBP curves.
   b. Stream gravities are obtained from the crude assay, and molecular weights are calculated.

Flash Zone and Tower Base Calculations

The following discussion outlines procedures for calculating the overall heat balance around the flash zone and tower base. These procedures apply for both types of vacuum towers and are considered independently of the rest of the heat and material balance calculations. To this point, it is assumed that the following items have been completed.

1. An overall material balance for the system has been developed including overflash and the amount of hydrocarbon decomposition gases which are produced in the process by virtue of the high temperatures to

Table 3.1

Pressure Drop Values As Recommended for Design Purposes	
Sections	Drop Values, e.a.
Chimney trays	
Fuels towers	6 to 8 mm Hg
Lube towers	1 to 2 mm Hg
Fractionating Trays	2 to 5 mm Hg
Grid sections	1 to 2 mm Hg
Demisters	1 mm Hg

which the reduced crude is subjected. The symbol, G, is used for this material in the calculations.

2. A value for inward air leakage is assumed which can be checked later by the methods outlined in Ludwig. For the heat balance, this leakage is assumed to occur totally at the flash zone. The symbol, A, is used for this material in the calculations.
3. The maximum allowable flash zone temperature has been established. As pointed out in Chapter 2, this has been the subject of much discussion throughout the industry. Normally, maximum flash zone temperatures range from 775 to 800 degrees F.
4. The tower top pressure has been established.

Flash Zone Pressure

In order to determine the flash zone pressure, it is first necessary to allocate trays and/or other internals to various sections of the tower and then to assume pressure drops across these sections. The number of trays between lube tower draw trays is normally 3 to 5. Modern fuels towers use grid sections. Pressure drop values recommended for design purposes are given in Table 3.1. Having assumed the internal configuration of the tower, the flash zone pressure is then calculated arithmetically.

Steam Requirements

Knowing the flash zone temperature and the feed vaporization requirements, the hydrocarbon partial pressure, P'_{HC}, in the flash zone is found by inspection of the vacuum region EFV curves which were developed earlier in this chapter. The difference between the partial pressure and the total pressure in the flash zone must be made up by the air leakage and by steam. The required steam to the flash zone is calculated as

$$S_W = [(D/y_{HC}) - D] - A, \text{ moles per hour}$$

where

D = total hydrocarbon distillate products including the decomposition gas, moles per hour, and

y_{HC} = ratio of hydrocarbon partial pressure to total pressure in the flash zone, mole fraction.

This analysis assumes essentially no stripping of the feed flash liquid in the base section of the tower, although, in a later step, a temperature drop across the bottoms stripping zone will be set. Nelsons' correlations show very little stripout at the steam rates usually employed, but experience shows that temperature drops as high as 30 degrees F between flash zone and bottoms do occur. This is of little real importance, however, since the heat input to the system can be calculated independently of the absolute thermal condition of the feed.

The error inherent in an incorrect estimate of the bottoms temperature carries two considerations.

1. From the viewpoint of the furnace, a low estimate of the exit bottoms temperature will cause the heat input to the system to be calculated as lower than actually required.
2. From the viewpoint of the tower heat balance, a low estimate of the exit bottoms temperature will cause the various tower heat removal quantities to be calculated as lower than actual. This applies especially to the cooling requirements for the vacuum residuum.

In practice, the value of these discrepancies is quite small in comparison to the total values of these heat quantities in question. The most conservative design approach would be to assume a zero temperature drop for the liquid from the flash zone to the base of the tower. This would maximize both feed heat input and heat removal duties. Conversely, assuming a high temperature drop, say 30 degrees F, would be a tighter, more competitive approach.

Heat Quantities

The calculations necessary to define the heat balance around the base section of the tower are shown on Figure 3.8. This illustration is self-explanatory as to the calculation of the total feed heat input and the furnace duty.

Note the role played by the condensed overflash liquid. If this stream is returned to the system, either as overflow from the draw tray to the flash zone or as recycle from the draw tray to the furnace, it must be taken into account in the heat balance. If it leaves the system as a product stream, it will not be seen as a heat input to the flash zone. This is the reason for the *question mark* in the equation for calculating the feed heat input.

The tabulation of the external heat quantities at the top of the flash zone is made to facilitate the heat balance

EXTERNAL HEAT QUANTITIES LEAVING FLASH ZONE

$Q'_{iFZ} = (Q_F + Q_A + Q_{SW})$, BTU/HR.

$Q'_{OFZ} = Q_W$, BTU/HR.

$\Delta Q'_{FZ} = (Q'_i - Q'_o)_{FZ}$, BTU/HR.

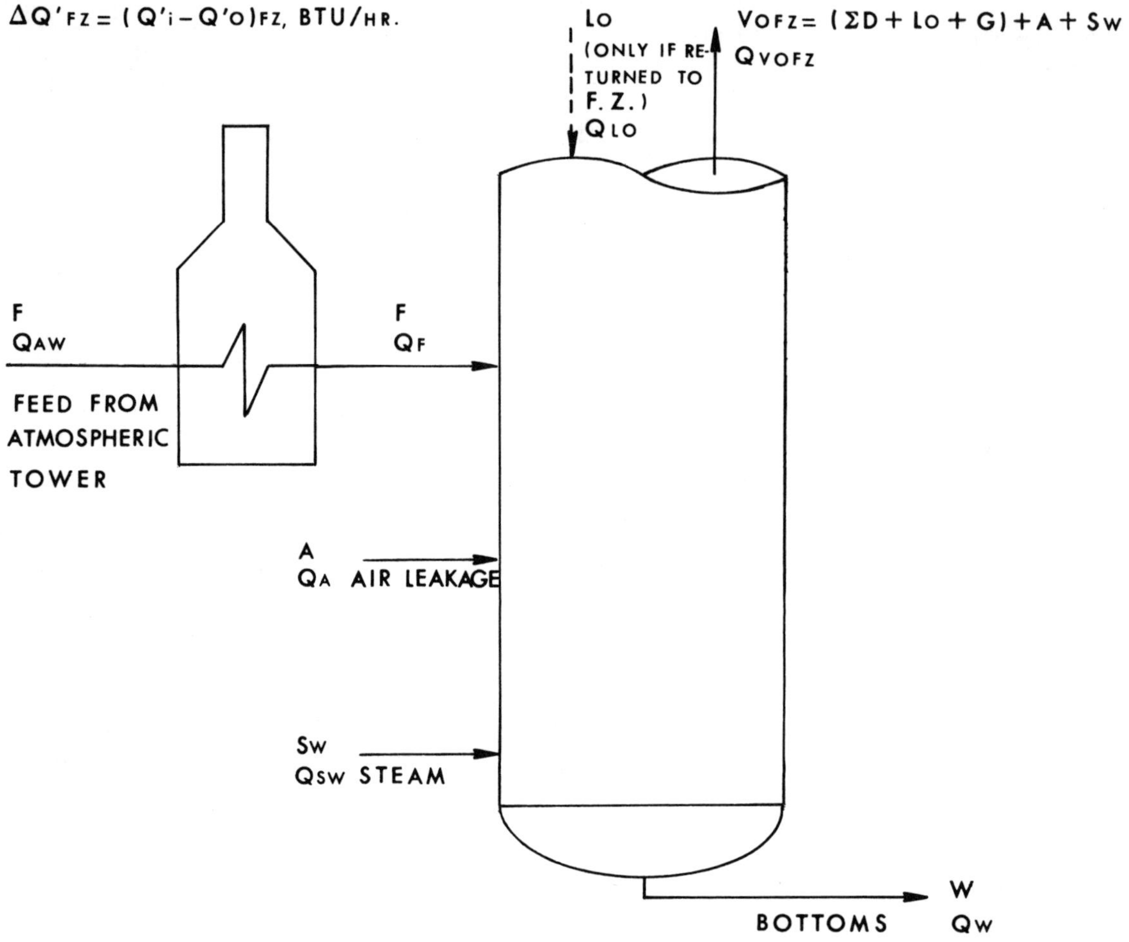

ΣD = TOTAL DISTILLATE PRODUCTS EXCLUSIVE OF OVERFLASH.

Lo = OVERFLASH.

G = HYDROCARBON DECOMPOSITION GAS.

HEAT BALANCE EQUATIONS

$Q_F = \left[(Q_{VOFZ} + Q_W) - (Q_A + Q_{SW} \overset{?}{+} Q_{LO})\right]$, BTU/HR.

$Q_{FURN} = (Q_F - Q_{AW})$, BTU/HR.

Figure 3.8. Flash zone heat and material balance.

calculations in proceeding up the tower. The value of the tabulation will be seen in later sections.

Heat and Material Balance Calculations for Lube-Type Towers

This section outlines procedures for calculating product draw tray temperatures at all points in the tower and for making an overall heat balance around the system. The method is based on calculating the hydrocarbon product partial pressure in the vapor above each draw tray and converting the bubble point of the product liquid on the tray to this pressure. Prior to beginning these calculations, the overall system material balance and the properties of all product streams must have been defined.

Steam to Product Strippers

These auxiliary towers will be used to strip only product and not the pumpback reflux, with the possible exception of the first product above the overflash liquid condensing section. For design purposes, set a steam rate of 10 pounds per barrel of final product as measured at 60 degrees F. This will result in a stripout of approximately 5 weight percent and a temperature drop across the stripper of 15 to 20 degrees F. Locate and plot the EFV temperature for this degree of stripout on the product EFV curves which were drawn in the calculations following the previous section. Note that the bubble point to be used in the partial pressure calculations is that of the unstripped liquid on the tray.

Knowing the thermal condition of the steam, calculate the heat input to all the strippers.

Tower Top Conditions

The temperature of the steam and noncondensible materials leaving the top of the tower is determined by setting a 50 to 75 degree F approach to the minimum practical cool oil temperature in the top pumparound system. This latter temperature is a function of the viscosity properties of the oil in question, and this data can usually be predicted from the crude assay. Normally, a cool oil temperature of 150 to 200 degrees F will not require excessive pump horsepower. This, in turn, allows an overhead temperature of 200 to 275 degrees F.

The amount of oil lost with the overhead stream is determined by making an oil vapor pressure calculation, assuming that the oil exerts its full vapor pressure at the temperature of the exit vapor. Maxwell's nomogram is useful in estimating this vapor pressure. The overall material balance should be adjusted to show the amount of oil lost with the overhead vapor. This oil loss is at the expense of the top sidestream product.

Calculate the amount of heat which leaves the system in the overhead vapor.

Estimate of Tower Temperature Profile

Estimating draw tray temperatures for vacuum operations is much more difficult than in atmospheric towers because of the greater relative effect of calculated internal reflux on hydrocarbon partial pressure. As good a rule as any is to assume a hydrocarbon partial pressure equal to 30 to 50 percent of the total pressure at any tray. Plot the assumed profile for the trayed section of the tower.

The temperature of the cooled pumpback reflux is estimated at 100 to 150 degrees F lower than the draw tray temperatures except where, for specific reasons, a different value would be indicated. Upon completion of the total design, it is necessary to verify that the heat exchange system is capable of producing the assumed reflux temperatures. If this analysis uncovers significant differences, it will be necessary to recalculate the tower, based on a different set of cooled reflux temperatures.

Overflash Liquid (OL) Condensing Section

The overflash liquid condensing section is calculated as a single-stage flash condensation. The heat removal across this portion of the tower is accomplished internally by revaporizing subcooled reflux which is pumped back from the next draw tray up in the tower.

Since this draw tray is a chimney tray and performs no fractionation, the temperature of the condensed liquid leaving the tray is estimated as being the dew point of the hydrocarbon vapor from the flash zone at the hydrocarbon partial pressure above the draw tray, the leakage air and the bottoms steam being defined as inerts. The temperature at this pressure is read from the vacuum region EFV curves. The vapor temperature leaving the grid is estimated in the same way but referred to the total pressure at that point.

The heat and material balance relationships around this section are determined by making a balance around Envelope I as shown on Figure 3.9. Figure 3.10 is an expanded view of this section and gives the equations used in making the calculations. These equations are to be used in the following sequence.

1. Calculate the reflux heat at the exit of the grid. Reflux heat is defined as the difference between the heat input of the feed and at the product strippers and the heat outflow of liquid products, external cooling of tower streams (product coolers excluded) and the exit vapors of products plus steam at the point in question.

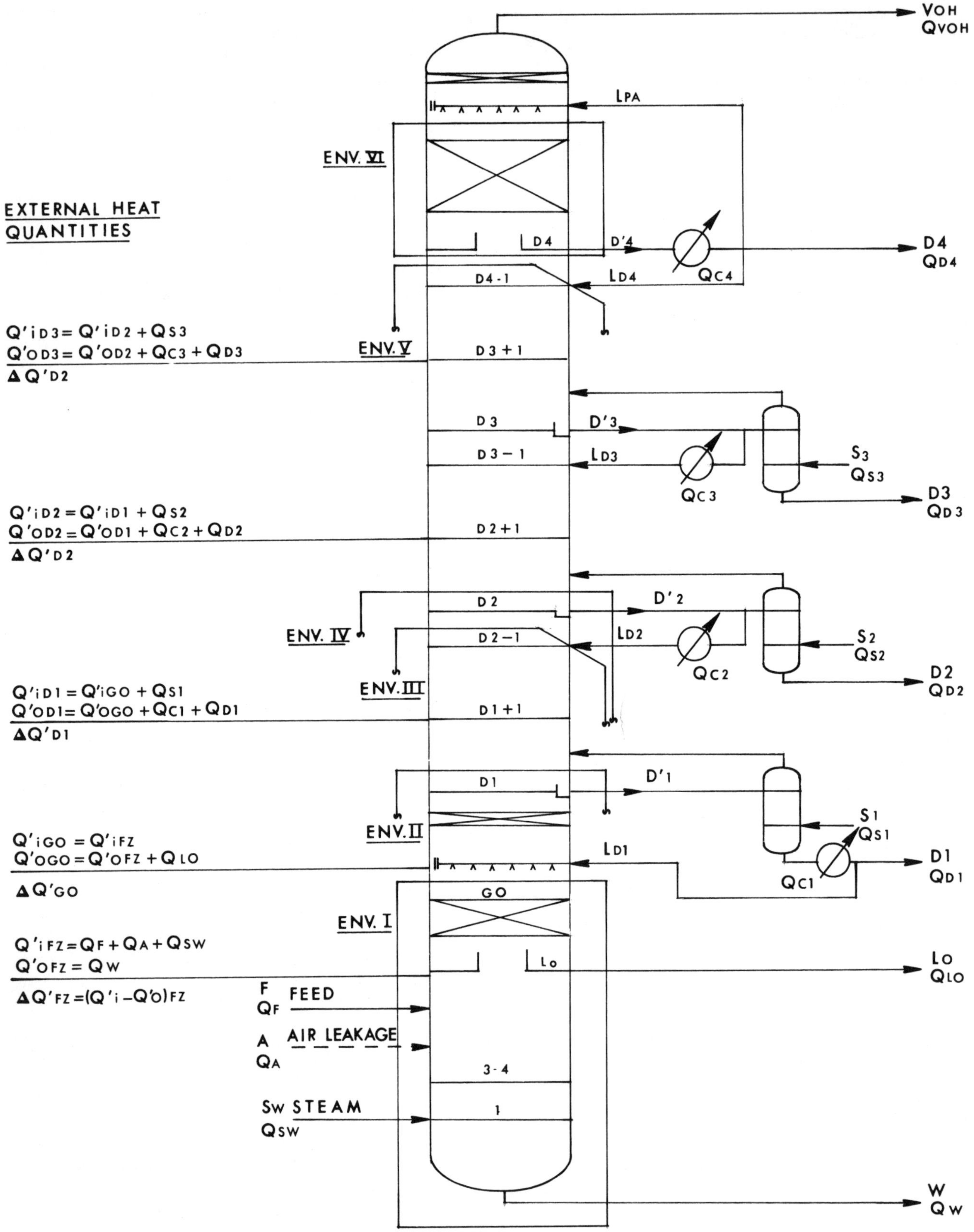

Figure 3.9. Heat and material balance summary, lube type vacuum tower with pumpback reflux heat removal.

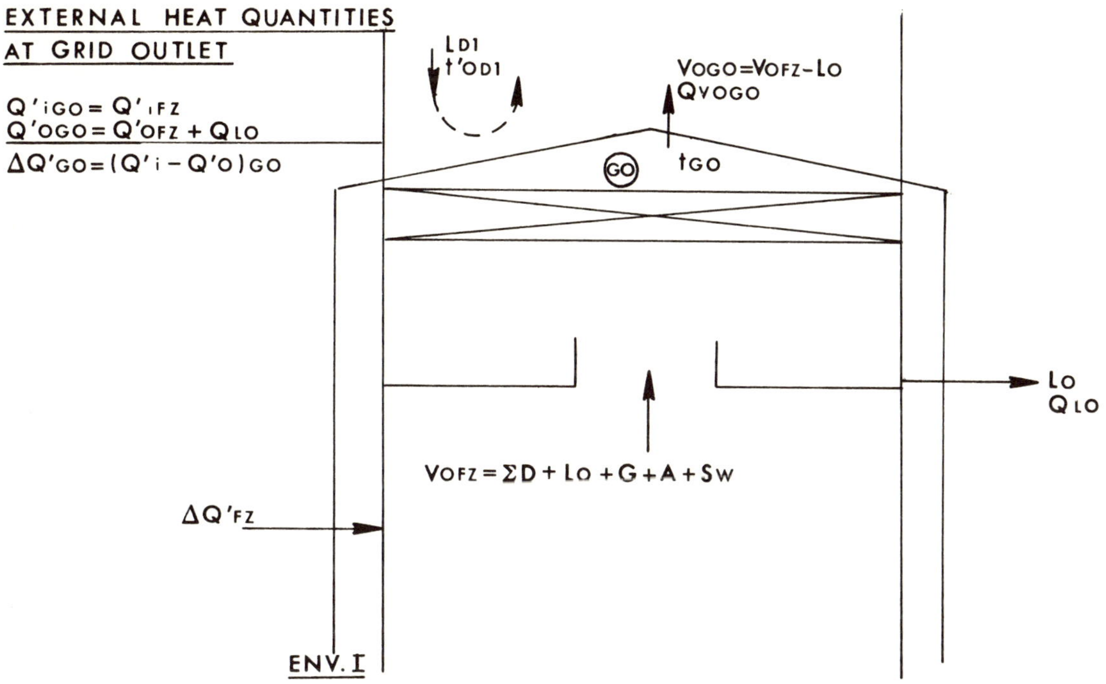

HEAT BALANCE EQUATIONS

REFLUX HEAT = $Q_R = (\Delta Q'_{FZ} - Q_{VOGO} - Q_{LO})$, Btu/hr.

HEAT REMOVAL CAPABILITY OF REFLUX $= q'_R = (H_{tGO} - h_{t'OD1})$, Btu/lb.

REQUIRED PUMPBACK REFLUX $= (Q_R/q'_R)$, lb./hr.

Figure 3.10. Heat and material balance—overflash liquid condensing section.

2. Calculate the heat removal capability of the reflux available to the tray.
3. Calculate the pumpback reflux required to absorb the reflux heat.
4. Calculate and tabulate the external heat quantities as Q′ values at the grid outlet.
5. Calculate and tabulate the vapor and liquids quantities leaving the grid.

First Sidestream Product (D1) Draw Tray

The heat and material balance relationships at this section of the tower are determined by making a balance around Envelope II as shown on Figure 3.9. Figure 3.11 shows an expanded view of this section and gives the equations used in making the calculations. These equations are to be used in the following sequence.

1. Calculate the reflux heat at Tray D1.
2. It is assumed that the amount of hydrocarbon which is revaporized in the D1 product stripper falls to Tray D1 as part of the internal reflux from Tray (D1 + 1) and absorbs a small amount of the reflux heat in passing across the tray. Calculate the value of this heat removal.
3. Calculate the heat removal capability of the internal reflux to the tray.
4. Calculate the internal reflux required to absorb the excess heat at Tray D1.
5. Convert the internal reflux from pounds per hour to moles per hour. Calculate the mole fraction of hydrocarbon product vapor in the total vapor leaving the draw tray but neglect the presence of the product to be removed on the next draw tray up in the tower. From this, calculate the hydrocarbon partial pressure in this vapor and convert the atmospheric bubble point of the unstripped liquid product on the tray to this partial pressure. If this temperature does not check the value assumed earlier, repeat the procedure for a new assumed temperature.
6. Calculate and tabulate the external heat quantities to the base of Tray (D1 + 1). This will include the heat effects of product liquid D1, the reflux cooler and the stripping steam to the product stripper.
7. Calculate and tabulate the vapor and liquid quantities at the base of Tray (D1 +1).

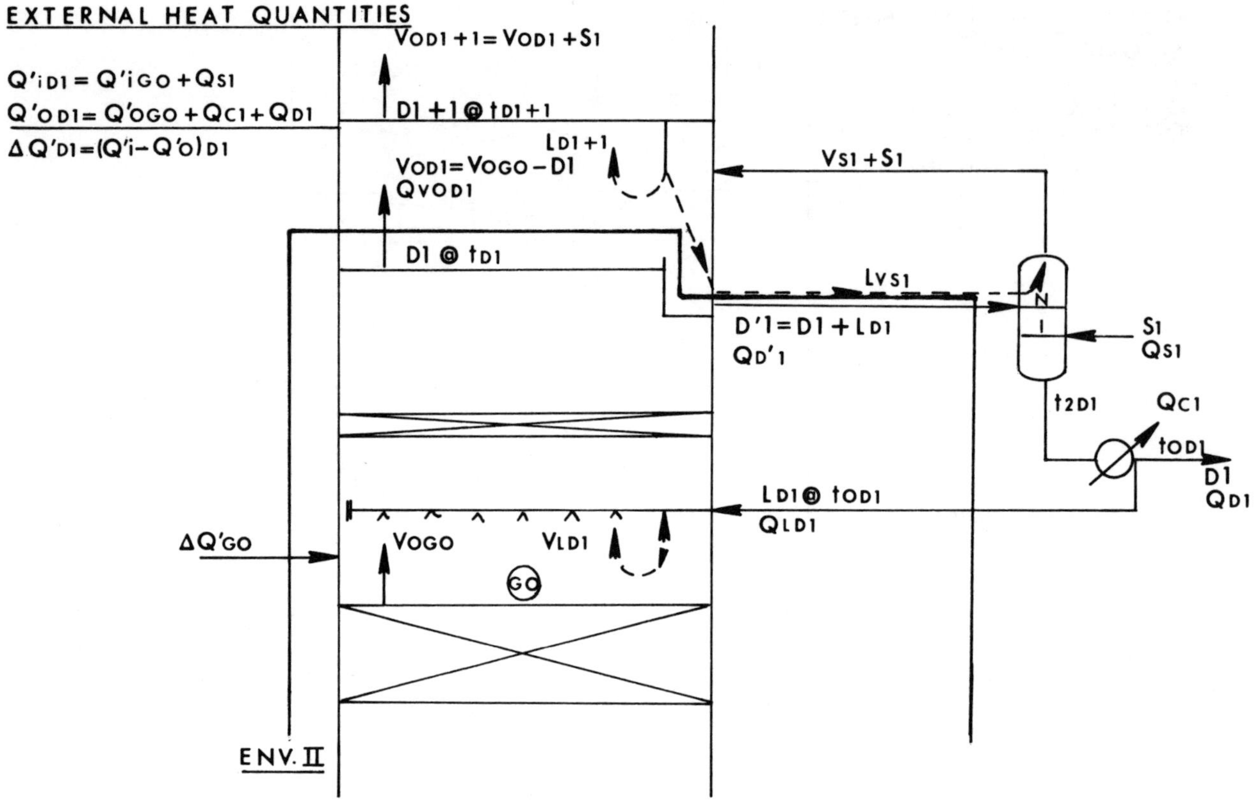

HEAT BALANCE EQUATIONS

$Q_R = (\Delta Q'_{GO} + Q_{LD1}) - (Q_{VOD1} + Q_{D'1})$.

HEAT ABSORBED BY $L_{VS1} = Q_{LVS1} = (L_{VS1})(\bar{c})(t_{D1} - t_{D1+1})$.

$q'_R = (H_{tD1} - h_{tD1+1})$.

$L_{D1+1} = (Q_R - Q_{LVS1})/q'_R$

HYDROCARBON PARTIAL PRESSURE ABOVE TRAY D1

$P'_{HC} = (P_{TD1})[L_{D1+1}/(V_{OD1} - D2 + L_{D1+1})]$

Figure 3.11. Heat and material balance—first sidestream product draw.

Second Sidestream Product (D2) Draw Tray

The heat and material balance relationships at this section of the tower are determined by making two balances. These balances are around Envelopes III and IV as shown on Figure 3.9 and in detail on Figure 3.12. The calculations for this section are similar to those in the previous section, with the exception of handling the pumpback reflux. The sequence of calculations is as follows.

1. Calculate the pumpback reflux required to Tray (D2-1) by making a balance above this tray as shown by Envelope III.
2. Calculate the internal reflux to Tray D2 by making a balance above this tray as shown by Envelope IV.
3. Calculate the hydrocarbon partial pressure in the total vapor leaving the draw tray, neglecting the presence of product D3. Convert the atmospheric bubble point of the unstripped liquid product on the tray to this partial pressure and check the assumed temperature.
4. Calculate the reflux induced on Tray (D2 −1) by the use of subcooled reflux. This induced reflux is the amount of vapor from Tray (D2 − 2) which enters and is condensed on Tray (D2 − 1) for the purpose of converting the cooled pumpback reflux to bubble point liquid.
5. Calculate and tabulate the external heat quantities to the base of Tray (D2 + 1).
6. Calculate and tabulate the vapor and liquid quantities at the base of Tray (D2 + 1).

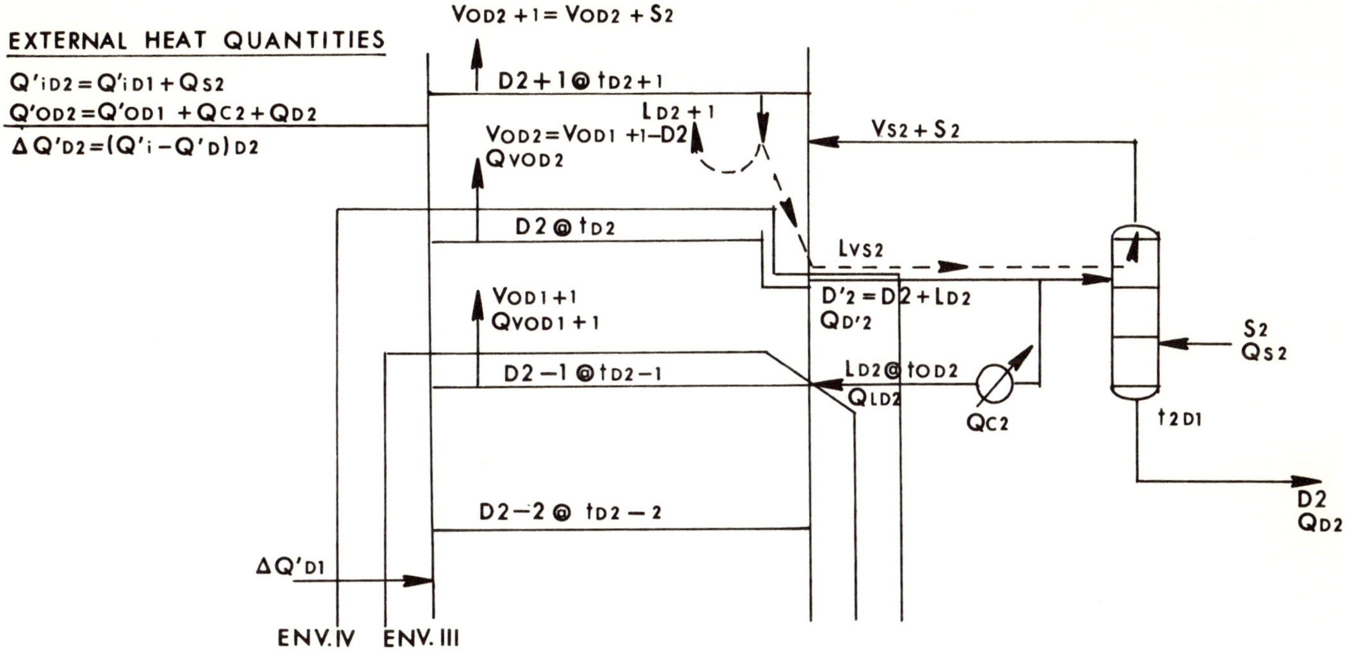

HEAT BALANCE EQUATIONS

(1) ABOVE TRAY D2−1 − ENV. III

$Q_R = [\Delta Q'_{D1} - Q_{VOD1+1}]$

$q'_R = (H_{tD2-1} - h_{tOD2})$

$L_{D2} = Q_R / q'_R$

ABOVE TRAY D2 − ENV. IV

$Q_R = [(\Delta Q'_{D1} + Q_{LD2}) - (Q_{D'2} + Q_{VOD2})]$

$Q_{LVS2} = (L_{VS2})(\bar{c})(t_{D2} - t_{D2+1})$

$q'_R = H_{tD2} - h_{tD2} - h_{tD2+1}$

$L_{D2+1} = (Q_{RD2} - Q_{LVS2}) / q'_R$

HYDROCARBON PARTIAL PRESSURE ABOVE TRAY D2

$P'_{HC} = (P_{TD2})[(L_{D2+1}) / (V_{OD2} - D3 + L_{D2+1})]$

INDUCED REFLUX ON TRAY D2−1

$R_{iD2-1} = L_{D2}[(h_{D2-1} - h_{tOD2}) / (H_{tD2-2} - h_{tD2-1})]$.

Figure 3.12. Heat and material balance—second sidestream product draw.

7. All further sections in the tower below the top sidestream draw tray are calculated in the same manner as Tray D2. Keep in mind that in making the partial pressure calculations, always neglect the presence of the next higher product vapor in the total vapor leaving the draw tray.

Top Sidestream Product (D4) Draw Tray

The analysis of this section of the tower involves calculating the required pumpback reflux to the tray below the draw tray together with the pumparound reflux and external heat removal requirements for condensing the sidestream product and cooling the tower overhead. The heat and material balance quantities around this section are shown by Envelopes V and VI on Figure 3.9. Figure 3.13 is an expanded view of this section and lists the equations used in the analysis. The recommended sequence of calculations is as follows.

1. A temperature on Tray (D4 − 1) was assumed in the preliminary steps. Using methods outlined previously, the required pumpback reflux is calculated and used to calculate the hydrocarbon partial pressure above the tray. In this case, the product vapor is defined as the sum of product D4 and the pumpback reflux while the

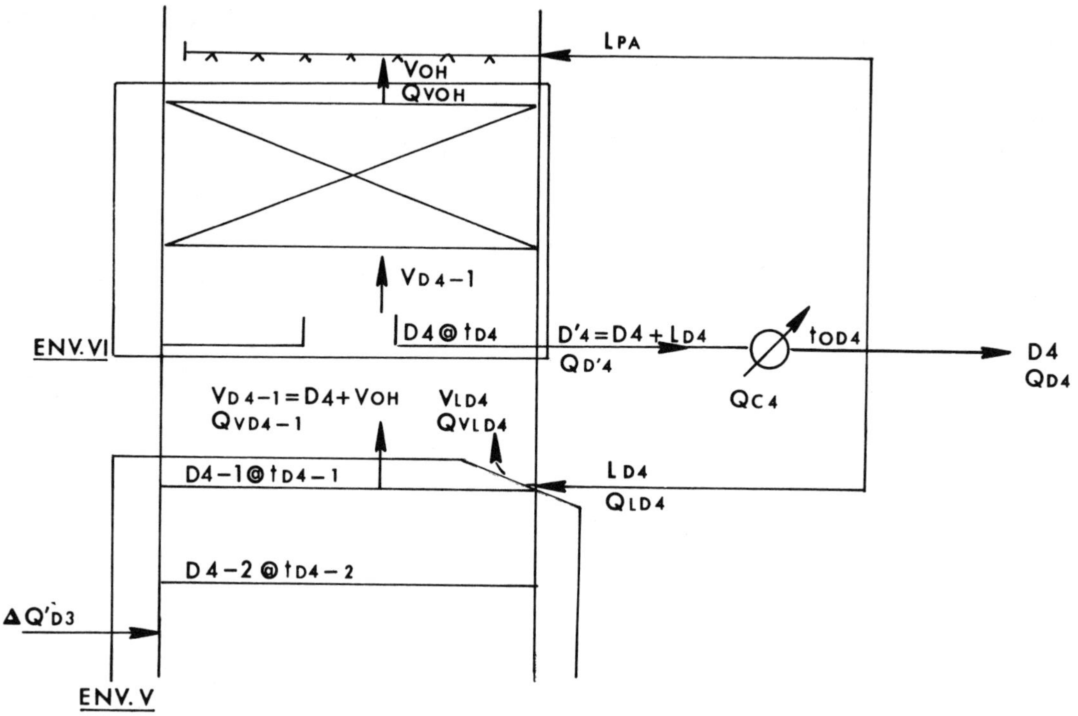

HEAT BALANCE EQUATIONS ABOVE TRAY (D4−1)−ENV.V

$Q_R = \Delta Q'_{D3} - Q_{VD4-1}$

$q'_R = H_{tD4-1} - h_{tOD4}$

$L_{D4} = Q_R/q'_R$

HYDROCARBON PARTIAL PRESSURE ABOVE TRAY (D4−1)

$P'_{HC} = (P_{TD4-1})[L_{D4}/(L_{D4}+D4+V_{OH})]$

HEAT BALANCE EQUATIONS AROUND D4 GRID

$Q_{C4} = (\Delta Q'_{D3} - Q_{D4} - Q_{VOH})$

$Q'_{C4} = D'_4 (h_{tD4} - h_{tOD4})$

$L_{PA} = (Q_{C4} - Q'_{C4})/(h_{tD4} - h_{tOD4})$

AS A CHECK, $(Q_{C4} - Q'_{C4}) = [(Q_{VD4-1} + Q_{VLD4}) - Q_{D'4} - Q_{VOH}]$

Figure 3.13. Heat and material balance—top sidestream product draw.

tower overhead stream comprises the inerts. Convert the atmospheric dew point of D4 to this hydrocarbon partial pressure and check the assumed temperature.

2. The temperature of the liquid on Tray D4 is found by converting the atmospheric bubble point of the product liquid to the hydrocarbon partial pressure existing above Tray D4.
3. Calculate the duty of the cooler as Q_{C4} by making an overall system heat balance, keeping in mind that the outlet temperature from this exchanger has already been defined.
4. Calculate the portion of Q_{C4} required to cool the liquid (D4 + L_{D4}) from the draw tray temperature to the exit temperature. Designate this quantity as Q'_{C4}. The remaining portion of Q_{C4} is that required to cool the pumparound reflux, L_{PA}.
5. Check the calculation of L_{PA} by making a heat balance around Envelope VI.
6. Check the value of Q_{C4} by calculating the heat requirements for cooling (L_{PA} + D4 + L_{D4}) over the required temperature range. This must check the value calculated in Step 4 above.

Vapor-Liquid Traffic

At all the trays which were calculated in the design, vapor and liquid flow rates were developed. In this section, these pieces of information will be assembled to develop an internal traffic diagram. At this point, the following items are known.

1. Vapor leaving the flash zone including total distillate products, overflash, steam, air and hydrocarbon decomposition gas.
2. Product streams leaving their respective draw trays.
3. Material balance around all product strippers.
4. Pumpback and pumparound reflux rates.
5. Internal liquid flow rates to draw trays.
6. Induced reflux where appropriate.

Using this information, developing the diagram reduces to a matter of arithmetic. The only place where caution must be used is in handling induced reflux. Consider a section of the tower containing the tray below a draw down to the tray above the next lowest draw. The known liquid rates in the section are pumpback reflux to the upper tray, its induced reflux and reflux from the lower tray. There is almost always a difference between the sum of the first two quantities and the third. In the analysis, this difference is assumed to be split equally between the trays in the section. This information is required for tray design or for analysis of the trays in existing towers. Discussion of tray design techniques is outside the scope of this work, and the interested reader is referred to the previously cited vendor literature.

Fractionation Analysis

There are no correlations known to the writer for analyzing fractionation in vacuum towers. The Packie analysis can be employed but only to show that a proper trays-reflux balance exists in the various sections. Estimates of gap-overlap may be made as a matter of interest, but it must be remembered that Packie's curves were intended to apply only to atmospheric crude towers.

Heat and Material Balance Calculations for Fuels-Type Towers

This section presents the remainder of the procedures required to calculate the heat and material balance around a fuels-type vacuum tower. Instructions for making the calculations at the flash zone and at the overflash liquid condensing section were presented in the previous section. As a quick refresher, the following items must be accomplished to this point.

1. Set the overall material balance, including air leakage and hydrocarbon decomposition gas.
2. Establish a pressure profile across the tower using previously recommended guidelines.
3. Make flash zone calculations of steam requirements and heat input to the tower in accordance with earlier sections.

Temperature Profile

Set the temperature of the vapor and liquid streams at key points in the tower. Unlike an atmospheric tower or a lube-type tower, these temperatures can be established analytically and do not require a trial-and-error approach. In this type of tower, the entire vapor charge to the tower is generated in the flash zone, and, with the exception of the overflash condensing section, there is no internal equilibrium reflux. All the heat is removed by pumparound reflux. Since the operation of the sidestream product condensing sections can be described closely as equilibrium condensations, the vapor temperatures can be estimated from the reduced crude EFV curves at the appropriate degree of vaporization and hydrocarbon partial pressure. Liquid temperatures are set by converting the atmospheric bubble points of the products to the appropriate partial pressures existing above the draw trays. This analysis also applies to the overflash liquid condensing section because the reflux required to condense the overflash is generally the same relative number of moles as the overflash.

The temperatures of the exit products and the pumparound reflux streams from their coolers are established taking into account the temperature-viscosity relationships of these very heavy oils. Product temperatures may also be set by heat balance considerations in downstream units.

Overflash Liquid Condensing Section

This section of the tower is calculated in the same manner as the lube-type tower. The procedure was discussed earlier and illustrated by Figures 3.19 and 3.10.

Sidestream Products (D1 and D2) Condensing Sections

The calculation of a fuels-type tower is much simpler than a lube-type. The material balance and heat balance relationships are analytical rather than trial-and-error in nature. Figure 3.14 shows the complete heat and material balance relationships for such a tower. The expressions and equations on this figure are self-explanatory. Note also that this sketch contains all the vapor-liquid internal traffic data.

References

1. D.E. Wheeler, "Design Criteria for Chimney Trays," *Hydrocarbon Processing* 47, no. 7 (July, 1968), pp. 119-20.
2. "Ballast Tray Design Manual," Bulletin No. 4900, Fritz W. Glitsch & Sons, Inc., Dallas, Texas.

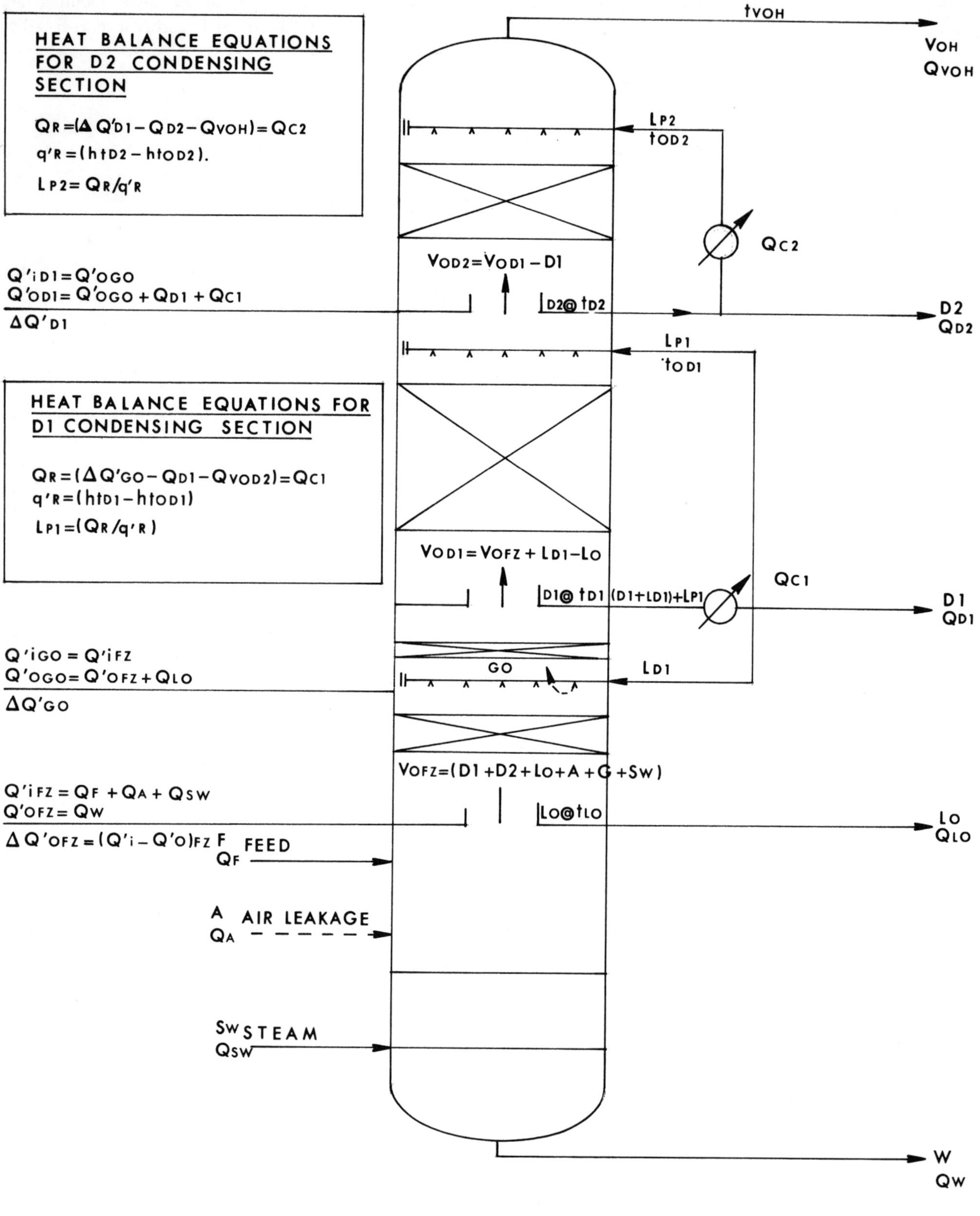

Figure 3.14. Heat and material balance summary—fuels type vacuum tower with pumparound heat removal.

3. *Koch Flexitray Design Manual*, Koch Engineering Co. Inc., Wichita, Kansas.
4. W.L. Nelson, *Petroleum Refinery Engineering*, 4th ed. (New York: McGraw-Hill Book Company, Inc., 1958).
5. "Glitsch Grid–A New Design for Column Internals," Bulletin No. 7070, Fritz W. Glitsch & Sons, Inc., Dallas, Texas.
6. E.E. Ludwig, *Applied Process Design for Chemical and Petrochemical Plants*, Vol. I (Houston: Gulf Publishing Company, 1964).
7. J.W. Packie, "Distillation Equipment in the Oil Refining Industry," *AIChE Transactions* 37 (1941), pp. 51-78.
8. W.C. Edmister, *Applied Hydrocarbon Thermodynamics* (Houston: Gulf Publishing Company, 1964).
9. J.B. Maxwell, *Data Book on Hydrocarbons* (Princeton, N.J.: D. van Nostrand Co., 1965).
10. F.W. Winn, "Physical Properties by Nomogram," *Petroleum Refiner* 36, no. 2 (February, 1957), p. 157.

main fractionator—catalytic cracking unit

The function of the catalytic cracking unit is to convert heavy straight-run distillate fractions into lighter and more useful liquids and gases by high temperature pyrolysis in the presence of a catalyst. Typical flow schemes and operating conditions are given in Reference 1. These reactions ordinarily occur at pressures of 10 to 15 psig and at temperatures in the range of 1,000 degrees F. The reactor effluent stream contains a full range of products from hydrogen through coke. At the entrance to the fractionator, it is a superheated vapor and must be cooled and simultaneously fractionated in carefully controlled stages to yield products having the desired properties.

The problems and calculational techniques involved in catalytic fractionator design are quite similar to those encountered in crude distillation, and, thus, the procedures given are somewhat less detailed to avoid useless repetition. The principal areas of difference between the two systems are as follows.

1. The feed to the tower is a superheated vapor, not a vapor-liquid mixture.
2. The heaviest materials must be condensed out and yielded as a bottoms liquid to prevent contamination of the first sidedraw liquid.
3. The sidedraw products usually have much wider boiling ranges than do the corresponding products from a crude unit. The properties of these streams are discussed in a later portion of this chapter.

The principal areas of similarity in the two systems are as follows.

1. The physical properties of the fluids being processed can be studied and calculated by the methods of Edmister (2) and Maxwell (3). The *API Data Book* (4) is also useful.
2. The interrelationship between trays, reflux and fractionation requirements are discussed in Reference 5 which presents a correlation similar to Packie (6). Later independent work has suggested that, for a more realistic approach, the systems should be calculated neglecting the presence of n-butane and all lighter components.
3. The use of pumparound reflux in these towers is almost universal. Pumpback reflux to the side sections of the tower is seldom used and, then, only in cases where a sharp separation between the cycle oils is required.

Keeping these points in mind, the tower is calculated in the same general way as is a crude tower.

Properties of Tower Feedstream (Reactor Effluent)

The composition of a catalytic cracking unit reactor product stream is usually described as a mixture of discrete light ends consisting of hydrogen through the C_5's, light and heavy cycle oils most often described by ASTM or TBP distillations and the heavy bottoms. The cycle oils may be broken down into subfractions identifiable by TBP boiling ranges. Alternately, the composition may be reported in the same way as a crude oil, i.e., a mixture of light ends (hydrogen through C_5's) and a TBP curve covering the range C_6 through end point.

If the first method for specifying the feed is used, the concentrations are usually reported on a weight basis, gravity data being available for the heavier cuts. This raw data is converted into a more usable form by the following procedures.

1. Using conventional physical property estimating methods (3,4), convert the weight units into volumetric and molar units.
2. The volumetric quantities are easily plotted into the familiar TBP-yield curve by graphical combination techniques.
3. Using Packie's procedure (6), convert the TBP curve into an atmospheric EFV curve.
4. Using Edmister's techniques, develop a phase diagram for the feed.

These procedures have all been discussed earlier in great detail and are mentioned here only to ensure that they are not forgotten.

A comparative study of catalytic fractionator feed and crude oil will show that they both have approximately the same TBP initial and final boiling points. However, the catalytic fractionator feed is much lighter in terms of TBP temperature versus volume percent vaporized. This is explained by the fact that the aromatics and condensed aromatics which make up the bulk of the catalytically cracked oils are much heavier than are the paraffins or naphthenes of similar boiling temperatures.

Product Properties and Separations

A catalytic fractionator can be designed to yield products of any desired boiling range in much the same manner as a crude tower. However, catalytically cracked fractions are generally subjected to further processing and/or treating before being sold and, thus, have a rather constant composition, regardless of refinery location. Most towers are designed to yield an overhead product–catalytic naphtha and lighter; two liquid sidestreams, commonly called light cycle oil, LCO, and heavy cycle oil, HCO; and a bottoms stream, commonly called slurry oil or decant oil. The overhead product stream further yields a vapor and a liquid, their compositions being determined by the temperature and pressure at which the equilibrium partial condensation occurs. In most units, the resultant vapor stream is compressed to a pressure level suitable for light ends recovery and is recombined with the overhead liquid stream, cooled and fed to a gas recovery plant. A typical catalytic cracking unit gas plant is shown in Figure 4.1.

Product Properties

Product streams from catalytic cracking unit main fractionators have the following general properties and end uses.

Catalytically Cracked Naphthas

This stream(s) consists of materials up to an ASTM end point of approximately 400 degrees F. Until relatively recently, stabilized catalytic naphtha was blended directly into motor fuel. In most modern refineries, however, this product is now desulfurized by mild hydrogen processing. Olefin saturation usually occurs also. Since catalytic naphtha is seldom used without further processing, it is not the usual practice to split the naphthas in the main fractionator. If the split should be required, it can be done easily. In any event, the tower and the reflux system should be designed to allow a reasonable variation in naphtha end point and sharpness of fractionation between the naphtha and light cycle oil.

Light Cycle Oil, LCO

Light cycle oil has the same general ASTM boiling range as the combined light and heavy atmospheric distillates–say 400 to 700 degrees F–and is used in the same general applications. If reasonably sulfur free, it can be sold directly as heating oil. It is commonly subjected to a mild hydrotreating for sulfur removal and for olefin saturation. It is often used as hydrocracking feedstock. It can be recycled back to the reactor along with the heavy cycle oil in order to increase conversion on the fresh feed. However, LCO is a poorer feedstock for catalytic cracking than is HCO because the former is rich in ring structures which are more difficult to crack. Thus, LCO recycle can be practiced only at the expense of conversion per pass. For these reasons, it is a more normal practice to yield LCO as a product stream from the unit.

Heavy Cycle Oil, HCO

Heavy cycle oil has the same general ASTM boiling range–say, 700 to 900 degrees F–as the combined atmospheric and vacuum gas oils yielded from a crude unit. This stream is almost invariably recycled to the reactor and cracked to extinction. Hence, facilities for cutting HCO into separate fractions are seldom required.

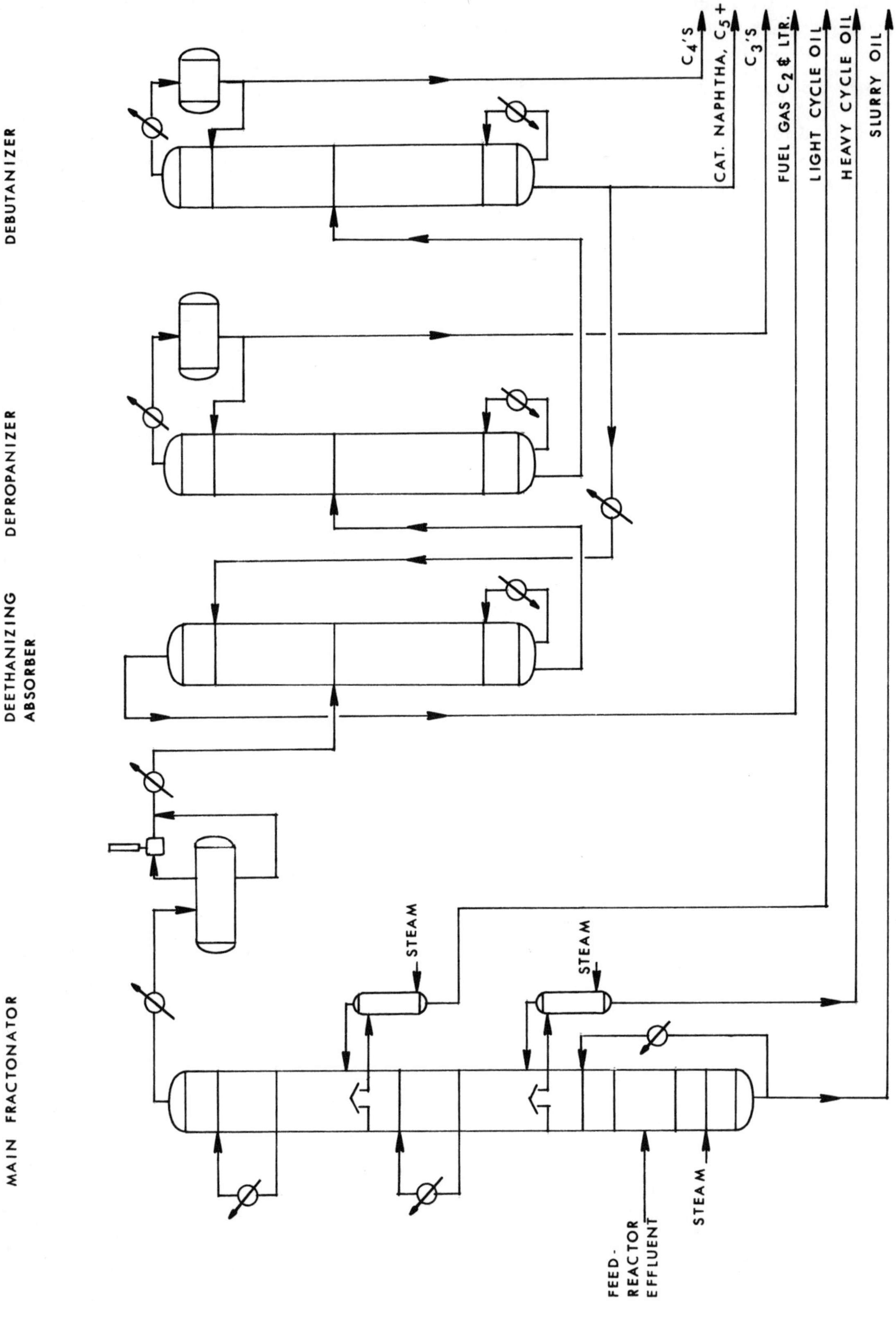

Figure 4.1. Catalytic cracking unit—product recovery section.

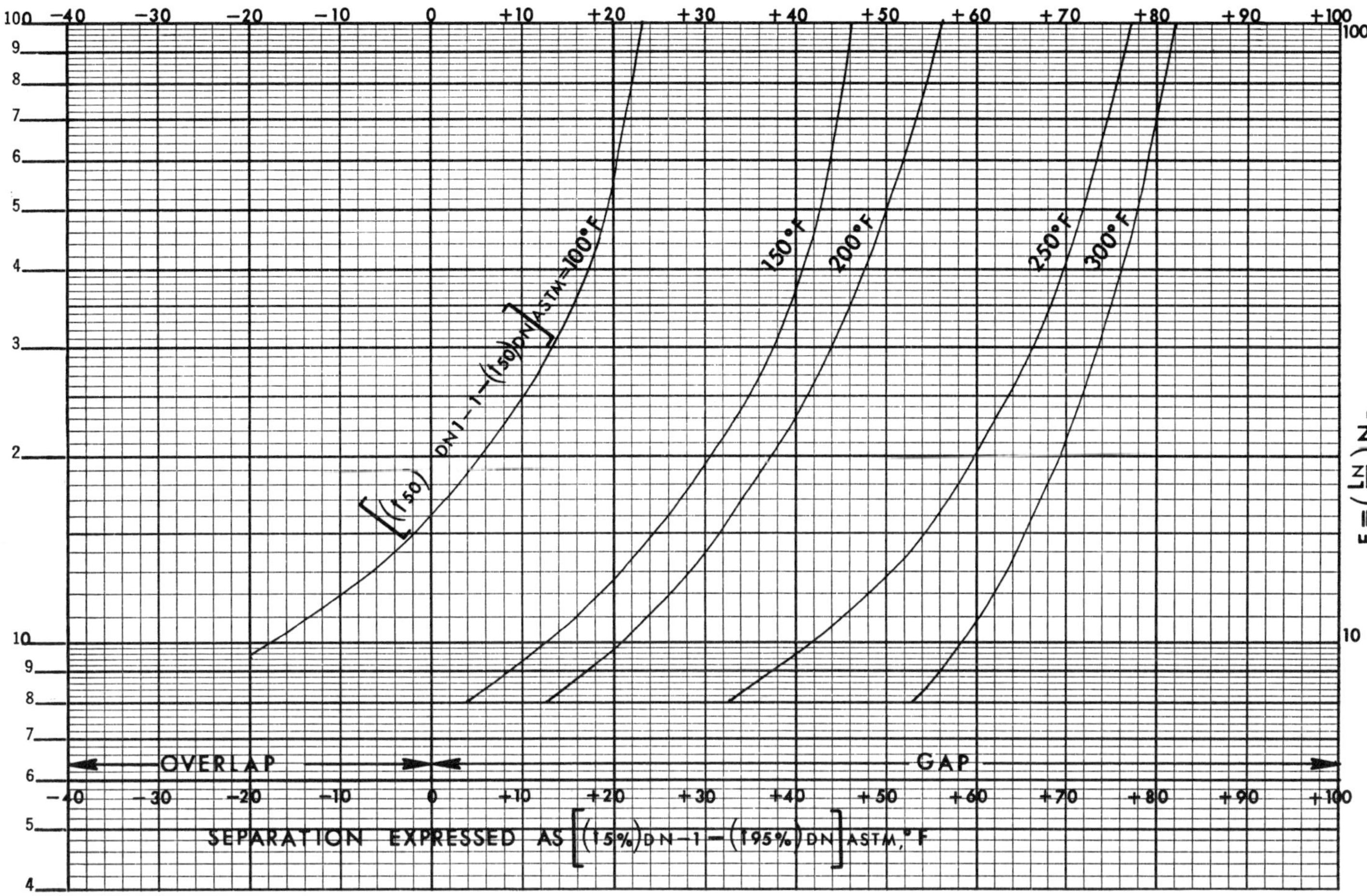

Figure 4.2. Fractionation between total overhead and highest sidestream product, catalytic cracking fractionators. (Used by permission of the American Petroleum Institute.)

Bottoms (Slurry Oil or Decant Oil)

The bottoms stream is actually nothing more than what is left over after the maximum distillate yield has been attained. This is a very foul stream and is usually blended into residual fuel. A rare but occasional practice is to design the base section of the tower as an HCO condensing section so that the bottoms streams will be HCO in addition to slurry oil. This operation requires that the bottoms slurry oil also be recycled to the reactor. This often results in low conversion to gasoline and more rapid coke formation.

Separation Criteria

As was the case in crude oil distillation, the sharpness of fractionation is described in terms of the ASTM (5 – 95) temperature differences. Packie's relationships, per se, do not apply to catalytic fractionation. However, Houghland, Lemieux, and Schreiner (5) have developed similar correlations for catalytic towers which are used in the same way as in which Packie is applied to atmospheric crude towers. Figures 4.2 and 4.3 have been drawn from this reference and are recommended for design and analysis of these units. The nomenclature for Figure 4.2 is as follows.

L_N = gallons per hour of equilibrium reflux from the top tray or gallons per hour equilibrium reflux from the bottom pumparound tray, measured as 60 degree F liquid.

D_N = gallons per hour total distillates to the top tray measured as 60 degree F liquid.

D_{N-1} = first sidestream product below top tray, measured as 60 degree F liquid.

N_T = number of actual trays in section. Note that each tray in pumparound heat removal service counts as one third of an actual tray. The nomenclature for Figure 4.3 is as follows.

L_N = gallons per hour reflux from the upper draw tray or gallons per hour reflux from bottom pumparound tray, measured as 60 degree F liquid.

P_N = gallons per hour total product vapors to upper draw tray, measured as 60 degree F liquid.

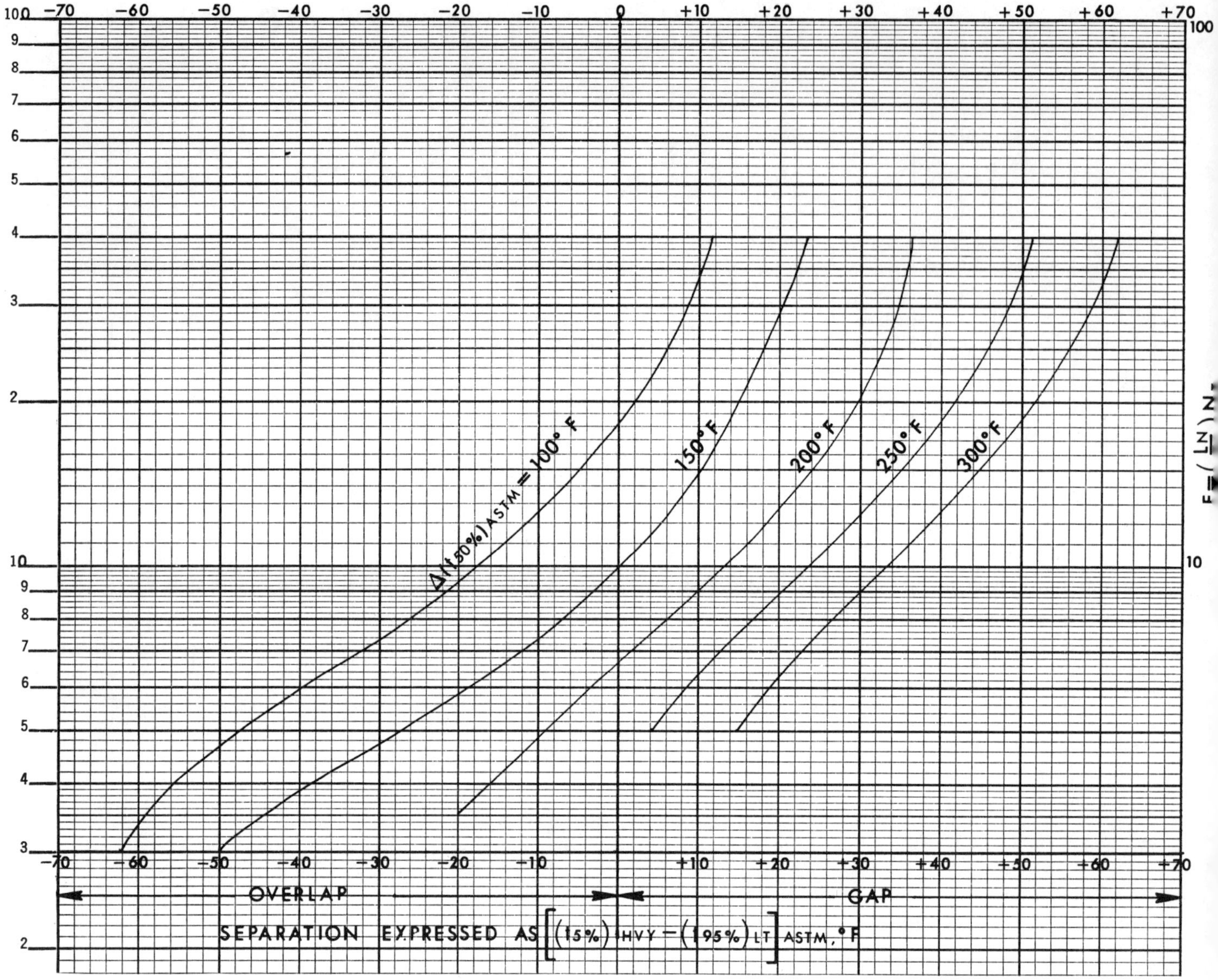

Figure 4.3. Fractionation between sidestream products, catalytic cracking fractionators. (Used by permission of the American Petroleum Institute.)

N_T = number of actual trays in section. Note that each tray in pumparound heat removal service counts as one third of an actual tray.

Both correlations require a steam stripping rate to product strippers of at least 8.4 pounds per barrel of stripped product. Subsequent information indicates that the correlation is more realistic if the presence of n-butane and all lighter components is disregarded in defining the product vapors. Thus, the effective internal reflux ratio from draw trays is greater than the apparent ratio since n-butane and all lighter components is disregarded in defining the product vapors. Thus, the effective internal reflux ratio from draw trays is greater than the apparent ratio since n-butane and lighter do not count in defining the product vapor. However, the original definition of the correlation is recommended since it represents a more conservative view of the problem.

Estimate of Material Balance

In a typical design, the material balance and separation criteria may be defined by:

1. Gross overhead ASTM end point.
2. ASTM boiling ranges for LCO and HCO.
3. TBP cut point between slurry oil and HCO.

4. ASTM (5 – 95) temperature differences may be specified or may be allowed to be the natural result of the above criteria.

These are translated into design information by the following stepwise procedure.

1. From the whole feed TBP curve, develop TBP cut points between fractions. Calculate and plot the ASTM curves of the naphtha and cycle oils. Figure 2.15 will be useful in converting between TBP and ASTM end and initial points.
2. Calculate and tabulate weight and molar yields of the fractions from which product gravities and molecular weights are calculated.
3. Calculate EFV curves for the overhead and for all sidestreams. Plot the EFV curves, extrapolating those for the sidestreams, LCO and HCO, to minus 20 percent.
4. Using Figure 2.13, set a steam rate to the sidestream strippers and estimate the material balance around these towers. For design purposes, a steam rate of 10 pounds per barrel of stripped liquid is recommended. At the indicated vaporization, find the EFV temperatures of the sidestreams at the appropriate minus vaporization percentages. These are the 14.7 psia bubble points of the product liquids on the draw trays and will be used later in calculating draw tray temperatures.
5. Set the steam rate to the tower bottoms at 10 pounds per barrel of bottoms product. In the case of catalytic towers and unlike crude unit towers, it is not necessary to set up a material balance around the feed entry point and the bottoms stripping trays because the vapor-liquid traffic at this point does not influence the overall heat and material balance calculations.
6. Tabulate the properties of the combined product vapors in the various sections of the tower. This will be very useful in making the heat and material balance calculations.
7. At the conditions of temperature and pressure existing in the overhead accumulator, calculate the vapor-liquid separation which will occur in this drum, remembering that water will exert its pure-component vapor pressure in the vapor phase. This procedure is exactly the same as that presented in Chapter 2.
8. After the number of trays and the draw tray locations have been set, plot the API gravity and molecular weight of liquid leaving trays versus tray number. Assume that liquids leaving the tray above and tray below draw trays have the same gravity and molecular weight as the product liquid.

Heat and Material Balance Calculations

The design of a catalytic fractionator is accomplished in essentially the same manner as an atmospheric crude tower. There is a slight difference in the techniques employed for analyzing fractionation capability, but, otherwise, the procedures are nearly identical.

Although the tower has the same process configuration as a Type A atmospheric crude tower, a different approach to setting tray and heat removal requirements has been taken here. Such an approach would easily lend itself to crude unit design and is of particular utility in revamp studies. In expansion studies, one is usually interested in maximizing throughput by increasing heat removal via pumparound. This procedure quickly shows how much heat is available for a given material balance.

Preliminary Steps

In getting the design underway, certain assumptions will be made and later checked against the results of the calculations.

Number of Trays

It is generally true that most towers in similar services will have about the same number of total trays and draw tray locations, regardless of where they are located geographically. Catalytic fractionators are no exception. The preliminary tray configuration of the tower is established based on the following general guidelines.

1. Trays are required between the feed entry point and the first product draw tray. These trays are for desuperheating the feed vapor to its dew point and for quenching and condensing that portion of the total feed which is to be yielded as bottoms product (slurry or decant oil). Depending upon the value of this heat quantity which must be removed, between four and eight trays will suffice. Because of the high tendency toward coke formation in this section of the tower, conventional trays such as valves or bubble-caps are not used here. Rather, trays having a high percentage of open area such as disc-and-donut, side-to-side baffles or "shed" angle iron decks are the types most commonly employed.
2. Four steam stripping trays are provided below the feed point for stripping recoverable material from the slurry oil. These trays are also of the high open area type.
3. In determining the design material balance and product properties, it is generally true that the separation requirements for the HCO-LCO separation are much less stringent than those for the LCO-naphtha separation. Accordingly, nine trays between the HCO and

LCO draw trays and 11 trays above the LCO draw tray are typical configurations. Each of these two sections will normally contain a three-tray pumparound heat removal section.

Definable Heat Quantities

At this early point in the design, the following heat quantities can be defined.

1. The feed temperature and pressure is given. Calculate the heat content of the feed. Note that the coke is treated as zero degree API liquid, both in the feed and in the bottoms product stream.
2. It is common practice to limit the temperature of the slurry oil to a maximum of 700 degrees F to minimize thermal decomposition in the section of the tower below the first product draw tray.
3. Assume a value for the temperature of the stripping steam. Calculate the heat content of all stripping steam streams.

Initial Premises for Process Design

It is necessary to make initial assumptions relative to proposed heat removal schemes prior to commencing design calculations. Since the separations made in catalytic fractionators are relatively easy in comparison with atmospheric crude tower separations and since the number of trays available is quite high by petroleum fractionation standards, it follows that internal reflux requirements are comparatively low. Thus, it is desirable to remove as much heat as possible from the system. Maximizing heat removal is usually accomplished by process-to-process exchange at various points in the unit, and this affords substantial utility savings. A secondary benefit is that internal reflux is minimized which, in turn, minimizes the tower diameter.

In normal practice, one should design for a high level of heat removal by pumparound systems. This is accomplished by setting the internal reflux from sidestream draw trays at zero or, at most, at very low values and employing zero reflux from the overhead condenser.

Many designers employ zero internal reflux from the lowest sidestream draw tray and accomplish the total heat duty in the lower section of the tower by pumparound heat removal. It has been the author's experience that a small amount of internal reflux to the lower section of the tower will do a more effective job of washing back the pyrolysis solids than does total heat removal by pumparound, even at very high liquid wash rates across the trays. This indicates that these pyrolysis solids are fractionated back to the bottoms rather than washed back by the scrubbing action of the liquid.

For the purposes of the illustrative design procedure, the pumparound heat removals were assumed on the following bases.

1. For the section of the tower up through the tray below the lower sidestream product (HCO) draw tray, let the pumparound heat removal take less than 100 percent of the total reflux heat, the remainder being satisfied by internal reflux from Tray A. A suggested value for design is 75 percent.
2. Set zero internal reflux from the upper sidestream draw tray, Tray B. A three-tray pumparound heat removal system is required for Trays (B – 1) to (B – 3).
3. Provide a pumparound heat removal system utilizing the top three trays in the tower. This heat removal will be such that no external reflux will be required from the condenser to the top tray. Another way of considering this is that the upper pumparound heat removal balances the system so that only overhead product leaves the top tray.

In using this scheme, check to see that the partial pressure of water in the overhead vapor is sufficiently low to preclude water condensation in the upper few trays of the tower. If water condensation does occur, it will be necessary to redesign the system to require some pumpback reflux from the condenser to the top tray. This will lower the partial pressure of water in the gross overhead vapor. Only enough pumpback reflux should be used to avoid the possibility of water condensation, the remainder of the heat removal being accomplished by the pumparound system.

Operating Pressure

The conditions of temperature and pressure of the feed at the tower inlet will almost invariably be set by the design of the reactor section. Assume a 5 psi pressure drop from the feed tray to the top of the tower and a 5 psi pressure drop from the top tray to the overhead product accumulator.

For later use in heat and material balance calculations, plot pressure versus tray number.

Vapor Temperatures to Draw Trays Assuming Zero Internal Reflux from Draw Trays

At this assumed operating condition, the vapor rising to draw trays will contain only products. Thus, the temperature of these vapors is the dew point of the particular product at the partial pressure at which it exists in the total vapor. As in the case of atmospheric and vacuum tower design, the presence of the product to be removed at the next draw tray up in the tower is neglected. All lighter product vapors and steam are considered as inert gases.

The temperature of vapor leaving the top tray is also calculated as the dew point of the overhead product—both vapor and liquid—at its partial pressure in this stream.

The flash zone temperature is determined from the phase diagram as the temperature of the total products leaving the feed section at the total hydrocarbon partial pressure existing there.

For later use in heat and material balance calculations, plot these temperature points versus tray number on the tray-pressure profile.

Process Flowsheet

Set up a process flowsheet such as Figure 4.4 which will be used in making the detailed heat and material balance calculations.

Heavy Cycle Oil (HCO) Draw Tray

The heat and material balance relationships at this section of the tower are determined by making balances around Envelopes I and II as shown on Figure 4.4. An expanded view of this section is given on Figure 4.5 together with the equations used in making the calculations. These equations are used in the following sequence.

1. Estimation of Lower Pumparound Heat Removal—Envelope I
   a. In an earlier step, the temperature of the vapor leaving Tray (A – 1) was calculated for the case of zero internal reflux from Tray A. At this temperature, calculate the heat content of the vapor leaving Tray (A – 1).
   b. From this and the heat quantities at points lower in the tower, calculate the duty of the lower pumparound cooler, Q_{C1}, which is required to satisfy the stipulation of zero reflux heat and, thus, zero internal reflux at Tray (A – 1).
   c. In accordance with the design premise, set Q_{C1} equal to the predetermined percentage of the valuc calculated above.
   d. Calculate and tabulate the external heat quantities above Tray (A – 1).
   e. Set the temperature of the cooled pumparound liquid at least 150 degrees F lower than the Tray (A – 1) temperature and calculate the lower pumparound circulation rate.
2. Tray A Balance—Envelope II
   a. Assume an operating temperature on Tray A. Calculate the heat content of the vapor and of the liquid product leaving Tray A at this temperature.
   b. From these heat quantities and the external heat quantities below Tray A, i.e., the external heat quantities above Tray (A – 1), compute the reflux heat above Tray A as Q_R.
   c. The hydrocarbon which is to be revaporized in the product stripper falls to the draw tray as part of the internal reflux from Tray (A + 1). In passing across the tray, it absorbs a small amount of the reflux heat. This heat quantity is defined as Q_{LSVHCO}.
   d. Calculate the heat removal capability of the reflux available to the tray as q'_R.
   e. Calculate the internal reflux required to absorb the excess heat at Tray A as L_{A+1} and convert it to moles per hour.
   f. Calculate the hydrocarbon partial pressure of L_{A+1} in the total vapor leaving Tray A. As in previous similar calculations, ignore the presence in the total vapor of the product to be removed on the next draw tray up in the column and consider all other lighter products and steam to be inerts.
   g. Convert the atmospheric bubble point of the unstripped hydrocarbon liquid on the draw tray to this partial pressure and check the assumed temperature. If there is a significant difference, assume a new temperature and repeat the procedure.
   h. Set the temperature of the stripped HCO from the sidestream stripper at 30 degrees F lower than the draw tray temperature and calculate the heat content of this product stream leaving the system.
   i. Calculate and tabulate the external heat quantities to the base of Tray (A + 1).
   j. Calculate and tabulate the vapor and liquid quantities above Tray A.

Light Cycle Oil (LCO) Draw Tray

The heat and material balance relationships at this section of the tower are determined by making balances around Envelopes III and IV as shown on Figure 4.4. An expanded view of this section together with the equations used in the computations is given by Figure 4.6. The following calculational sequence is recommended.

1. Calculation of Midpumparound Heat Removal—Envelope III
   a. In an earlier step, the vapor temperature leaving Tray (B – 1) was calculated for the condition of zero internal reflux from Tray B. At this temperature, calculate the heat content of the vapor leaving Tray (B – 1).

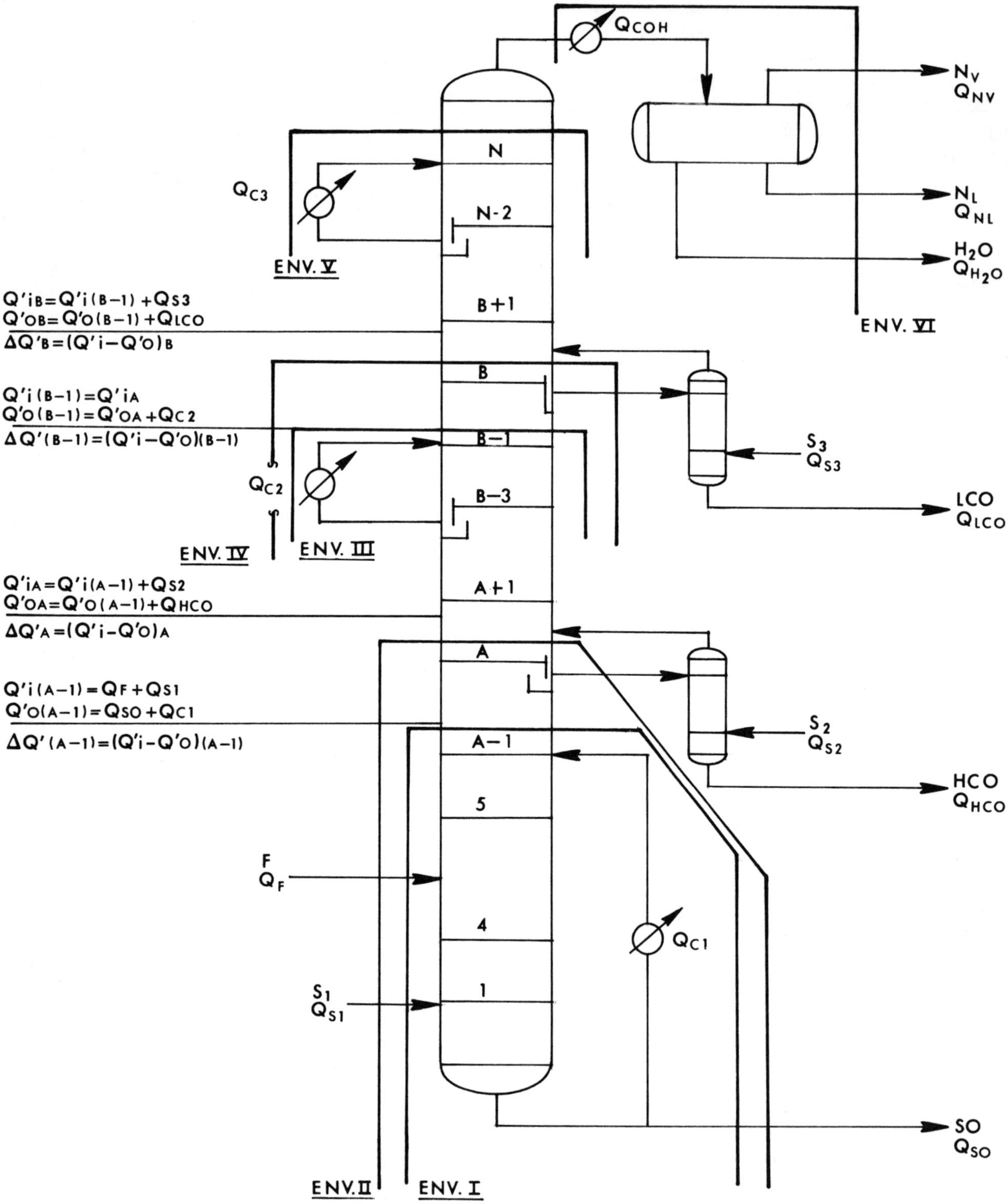

Figure 4.4. Heat and material balance summary, catalytic cracking unit—main fractionator.

EXTERNAL HEAT QUANTITIES

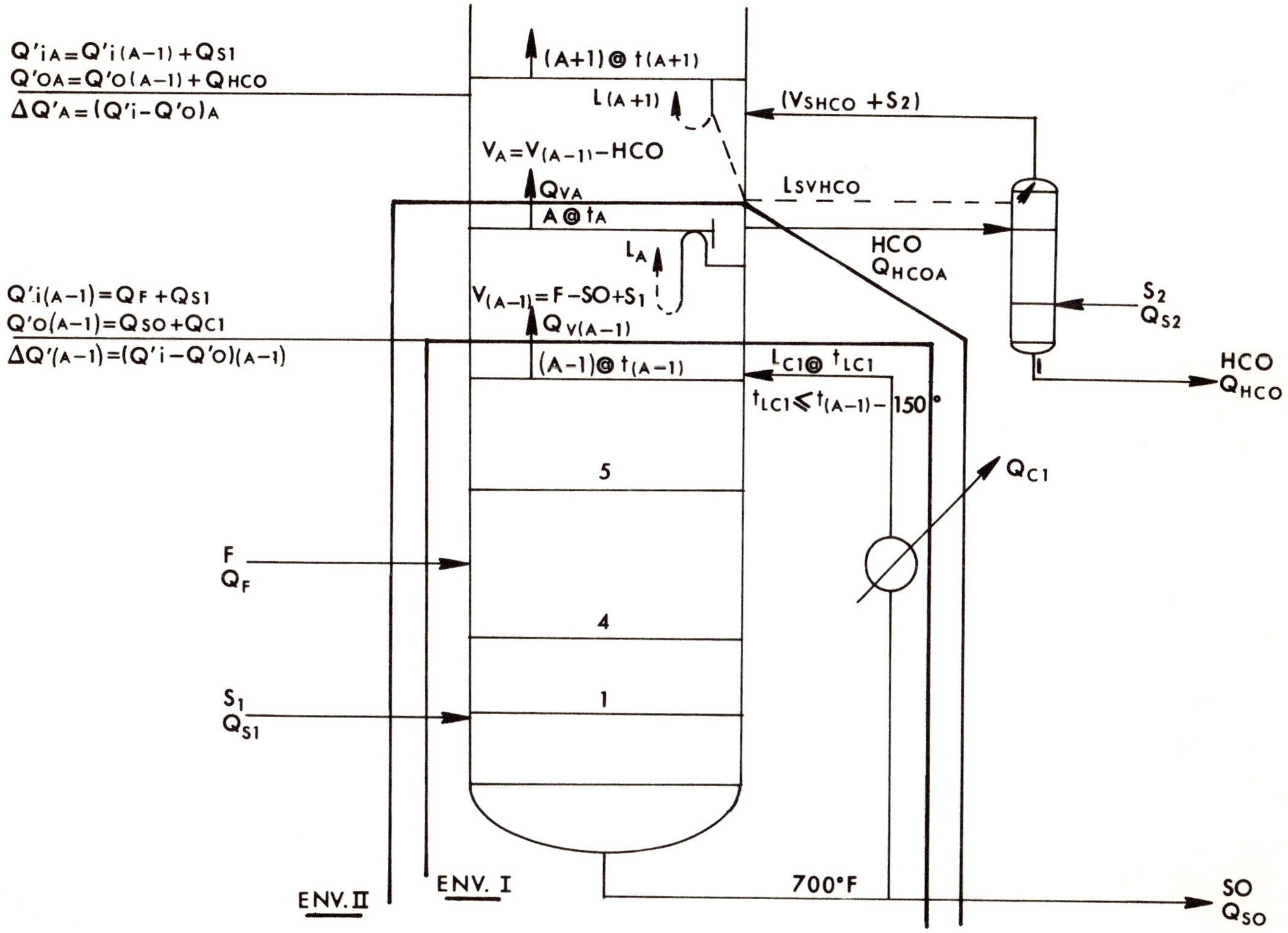

HEAT BALANCE EQUATIONS

(1) SETTING $L_A = 0, Q_{C1} = (Q_F + Q_{S1}) - (Q_{SO} + Q_{V(A-1)})$. } ENV. II

(2) DEFINE $Q_{C1DSGN} = .75\,Q_{C1}$

CALCULATIONS ABOVE TRAY A-ENVELOPE II

$Q_R = \Delta Q'_{(A-1)} - (Q_{VA} + Q_{HCOA})$

$Q_{LSVHCO} = (L_{SVHCO})(\bar{c})(t_A - t_{(A-1)})$

$q'_R = (H_{tA} - h_{t(A+1)})$

$L_{(A+1)} = (Q_R - Q_{LSVHCO})/q'_R$

PUMPAROUND CIRCULATION RATE

$L_{C1} = Q_{C1}/(\bar{c})(700 - t_{LC1})$

HYDROCARBON PARTIAL PRESSURE ABOVE TRAY A

$P'_{HC} = (P_{TA})\left[(L_{(A+1)})/(L_{(A+1)}) + V_A - LCO)\right]$

Figure 4.5. Heat and material balance quantities—HCO draw tray.

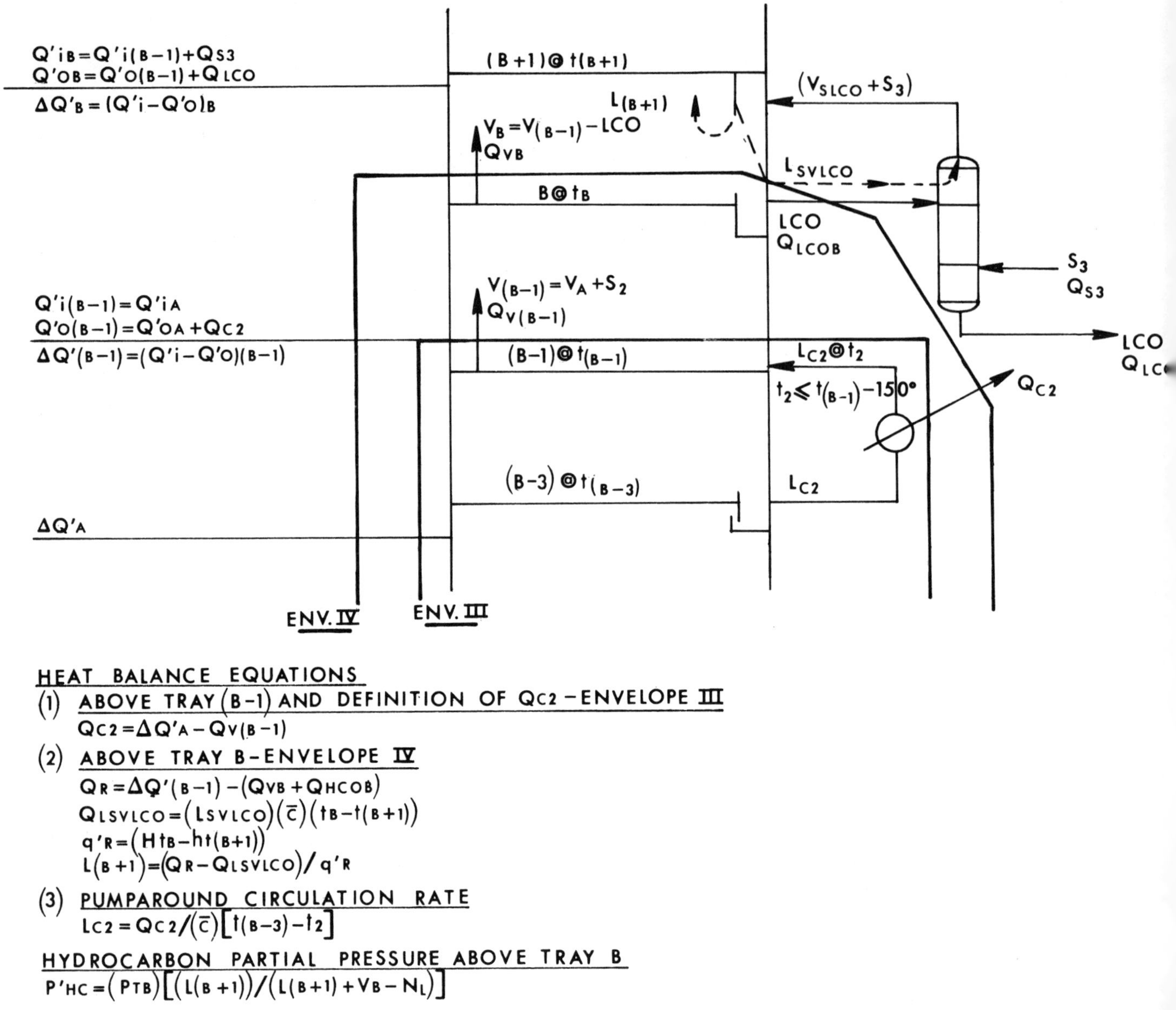

Figure 4.6. Heat and material balance quantities—LCO draw tray.

b. Calculate the heat removal required in the midpumparound cooler to produce zero internal reflux from Tray B.
c. Calculate and tabulate the external heat quantities above Tray (B − 1).
d. Set the cooled pumparound liquid temperature at least 150 degrees lower than the Tray (B − 1) temperature and calculate the midpumparound circulation rate.

2. Tray B Balance—Envelope IV
   a. At the assumed temperature, calculate the heat contents of the vapor and liquid streams leaving Tray B. Note that the liquid is to be considered as *only* the product quantity.
   b. Calculate the reflux heat above Tray B.
   c. Calculate the heat absorbed in passing across Tray B by the liquid which is to be revaporized in the sidestream stripper.
   d. Calculate the heat removal capability of the reflux available to Tray B.
   e. Calculate the internal reflux from Tray (B + 1) and convert it to moles per hour.
   f. Calculate the partial pressure of product vapor in the total vapor rising from Tray B, neglecting the

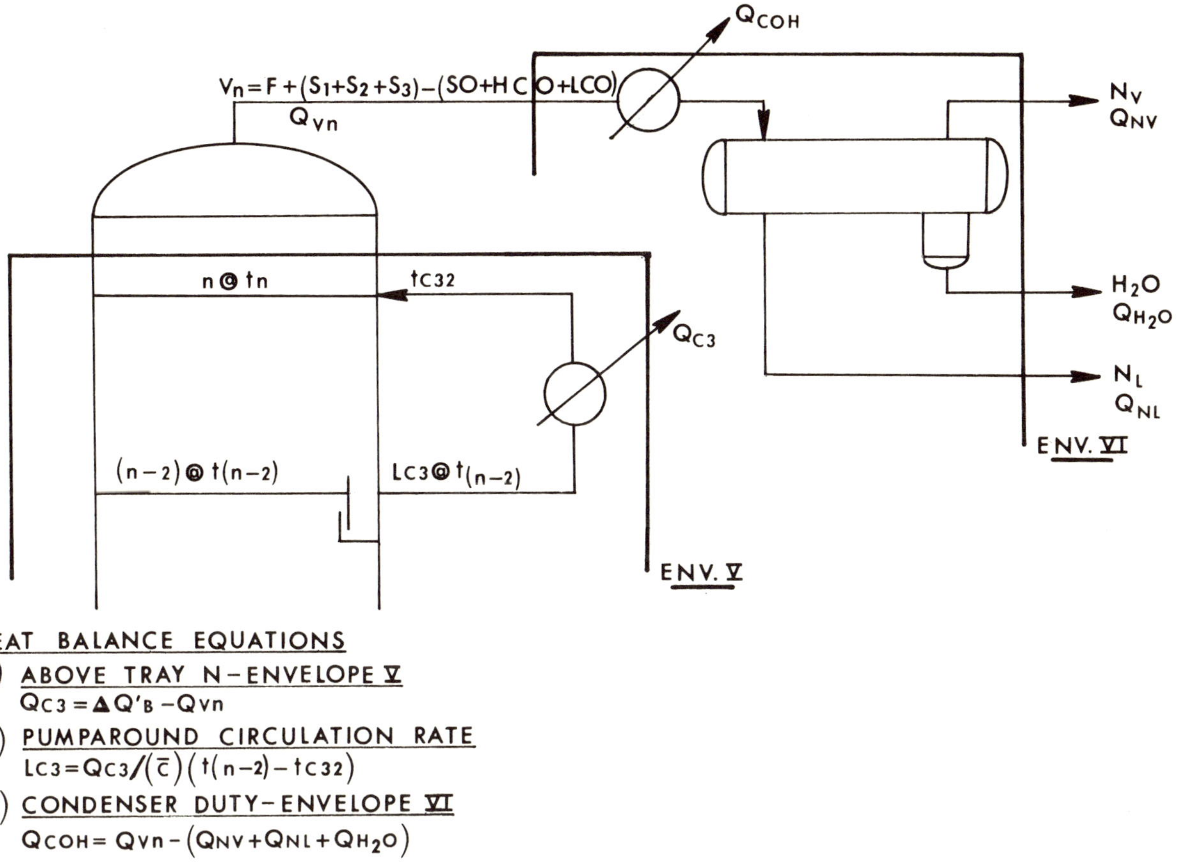

Figure 4.7. Heat and material balance equations—top tray and condenser.

presence in the vapor of the overhead distillate liquid and considering the overhead distillate vapor and steam as inerts.

g. Convert the bubble point of the unstripped hydrocarbon liquid on the draw tray to this partial pressure and check the assumed temperature.
h. Set the temperature drop across the sidestream stripper at 30 degrees F and calculate the heat content of the product leaving the system.
i. Calculate and tabulate the external heat quantities to the base of Tray (B + 1).
j. Calculate and tabulate the vapor and liquid quantities above Tray B.

Top Tray Calculations and Overhead Condenser Duty

The heat and material balance relationships at this section of the tower are determined by making balances around Envelopes V and VI as shown on Figure 4.4. An expanded view of this section together with the equations to be used in the computations is given by Figure 4.7.

1. Calculations of Top Pumparound Heat Removal—Envelope V
   a. In Step D, the temperature of the overhead product vapor was calculated on the basis that there would be zero external reflux to the tower from the condenser. At this temperature, calculate the heat content of the vapor leaving the tower. Note that this stream consists only of product materials and stripping steam.
   b. The top pumparound heat removal is now calculated as that amount required to balance the tower.
   c. In practice it may not be feasible to set the temperature of the cool pumparound liquid 150 degrees F lower than the overhead vapor temperature. In this case, take a reasonable approach to the temperature of the available cooling medium and calculate the top pumparound circulation rate.
2. Calculation of Overhead Condenser Duty—Envelope VI
   The overhead condenser duty is calculated as the enthalpy difference between the overhead vapor and the products from the overhead accumulator.

Fractionation Calculations

1. Calculation of Internal Reflux to Key Trays
   By making heat balances, calculate the internal reflux at the following points in the tower.
   a. Liquid to the top tray in the bottom pumparound section, Tray (A – 1).
   b. Liquid from the bottom tray in the midpumparound section, Tray (B – 3).
   c. Liquid from the bottom tray in the top pumparound section, Tray (N – 2). L_{B-3} and L_{N-2} are the liquid rates which are used in the fractionation analysis.
2. Fractionation Calculations
   Using Figures 4.2 and 4.3, determine the degree of separation possible for the system as calculated. If the fractionation criteria have not been satisifed, additional trays or reduced heat removal may be employed to achieve the desired separation.

Vapor-Liquid Traffic

Tabulate and plot the vapor and liquid traffic at key points in the tower. This plot will usually identify the points of maximum load in the tower and will be of great assistance in tower sizing and tray design calculations.

Process Design Considerations

Assuming that the heat and material balance calculations have been finished and that the design appears feasible, the remaining tasks are to ensure that the assumed pumparound configuration can indeed remove the required amount of heat from the tower.

Neeld and O'Bara (7) and Fair (8) have published procedures for calculating the heat transfer capabilities of "jet trays" and side-to-side trays, respectively. Jet trays are similar to conventional trays, e.g., bubble-cap or valve trays, which would normally be specified for this service, and, thus, Neeld and O'Bara's correlations may be used as checks.

Since tower diameters are used in the heat transfer calculations, tray sizing calculations may be made using procedures from Glitsch (9) or Koch (10) or any other method which the reader might prefer.

If an insufficient number of heat transfer trays were provided in the original design assumptions, they can be added as required with only minor modifications to the calculations.

References

1. "1970 Refining Processes Handbook," *Hydrocarbon Processing* (September, 1970), pp. 174-79.
2. W.C. Edmister, *Applied Hydrocarbon Thermodynamics* (Houston: Gulf Publishing Company, 1961).
3. J.B. Maxwell, *Data Book on Hydrocarbons* (Princeton, N.J.: D. van Nostrand Co., 1965).
4. *Technical Data Book–Petroleum Refining* (Washington, D.C.: American Petroleum Institute, 1966).
5. G.S. Houghland, E.J. Lemieux, and W.C. Schreiner, "The Performance of Catalytic Cracking Unit Fractionating Towers," 19th Midyear Meeting, API Division of Refining, (May 13, 1954).
6. J.W. Packie, "Distillation Equipment in the Oil Refining Industry," *AIChE Transactions* 37 (1941), pp. 51-78.
7. R.K. Neeld and J.T. O'Bara, "Jet Trays in Heat Transfer Service," *CEP 66*, no 7 (July, 1970), pp. 53-59.
8. J.R. Fair, "Design of Direct-Contact Gas Coolers," *Petro/Chemical Engineer* (August, 1961).
9. "Ballast Tray Design Manual," Bulletin No. 4900, Fritz W. Glitsch & Sons Inc., Dallas, Texas.
10. "Koch Flexitray Design Manual," Koch Engineering Co., Inc., Wichita, Kansas.

refinery light ends fractionation

Up to this point, this work has considered gross or rough separations. In the case of the atmospheric crude tower, petroleum was separated into relatively narrow fractions while, in the case of the FCCU main fractionator, the tower produced fewer fractions having wider boiling ranges. In both cases, the lightest (overhead) fraction was a full-range naphtha, i.e., everything in the tower feed up to a certain predetermined end point. This section will concern itself with the distillation processes required to separate light hydrocarbon components from the heavier continuous-boiling fractions, to fractionate the discrete light ends components and to separate the heavier continuous boiling materials into two or more fractions.

In the refinery, the term "light ends" generally means any discrete component lighter than heptane which can be identified by a name. This includes everything from hydrogen through the hexanes. A more narrow definition might consider only C_3 and C_4 liquids as light ends since, in many refineries, ethane and lighter is used as fuel gas and the pentanes and hexanes are blended directly into gasoline. As will be seen in the development of this section, there are many reasons for recovering light ends, each being dictated by the process configuration and economics of the refinery in question.

For a moment, let us explore the historical development of light ends recovery plants within the context of the total refinery. In the early days of the industry, kerosene, stove oil and lubricants were the principal products. With the advent of the automobile, gasoline became valuable and was produced. In most cases, the raw naphtha distillate was stored in open tankage where the light components vaporized and the gasoline became self-stabilized. Besides being wasteful, this practice was dangerous.

As the automobile engine developed, gasoline specifications became more stringent, and a need developed for better control of vapor pressure which radically affects carburetion and ignition. The refiner accomplished this by installing a naphtha stabilizer with which he could closely control gasoline vapor pressure. A secondary benefit was that the vapor distillate from this tower—essentially butanes and lighter—could be used as fuel gas within the refinery. Figure 5.1 shows a typical whole naphtha stabilizer producing a vapor distillate to the refinery fuel gas system.

In time, some enterprising person invented a stove utilizing as fuel either propane or butane, which the refinery could easily produce. It required that the naphtha stabilizer be redesigned at a higher pressure level in order to yield a liquid distillate which, in turn, could be fractionated into propane and butane as liquefied petroleum gas (LPG). Figure 5.2 shows a typical scheme for processing the whole naphtha stabilizer liquid distillate into propane and butane while still yielding fuel gas consisting of ethane and lighter.

As time passes and the petroleum industry and society develops, the demand for refinery products skyrockets, and product specifications continually tighten. New processes are developed to improve gasoline octane. Petrochemical

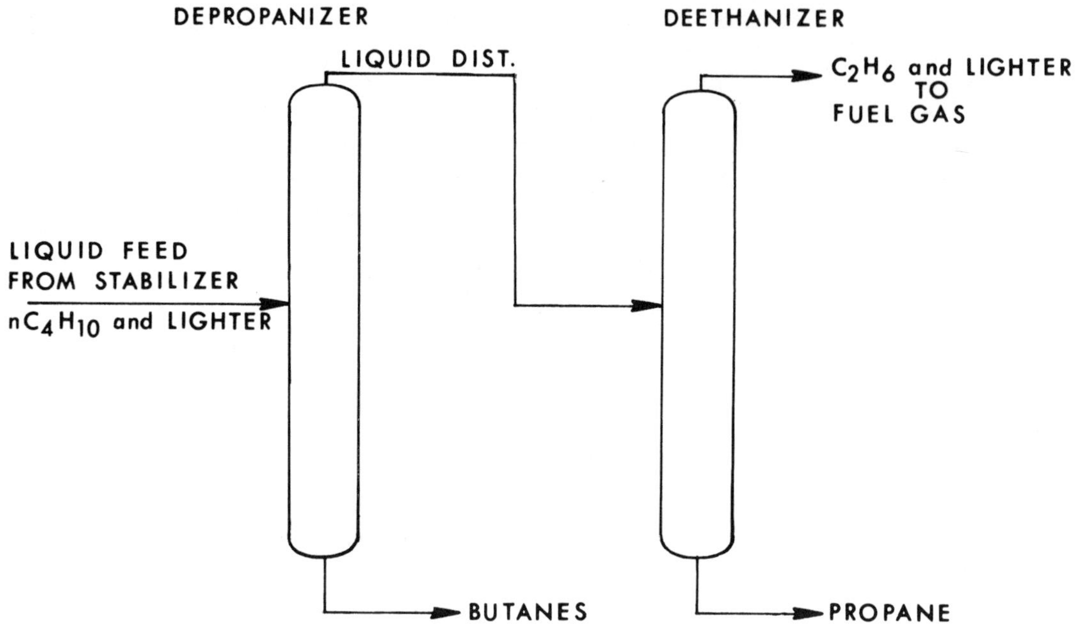

Figure 5.2. LPG recovery from refinery gas.

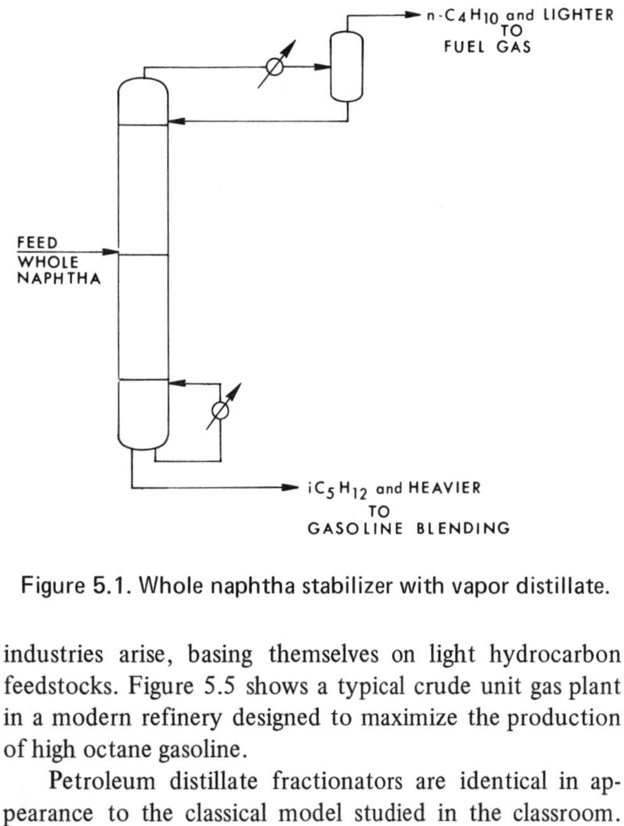

Figure 5.1. Whole naphtha stabilizer with vapor distillate.

industries arise, basing themselves on light hydrocarbon feedstocks. Figure 5.5 shows a typical crude unit gas plant in a modern refinery designed to maximize the production of high octane gasoline.

Petroleum distillate fractionators are identical in appearance to the classical model studied in the classroom. They are fitted with condensers and reboilers and process generally one feed into two or, at the most, three products. The feed enters the tower at an intermediate tray so that there is always a stripping section below the feed tray and a rectifying section above it. Unfortunately, this is the end of the similarity between petroleum fractionators and the classical distillation column normally studied in the classroom.

Consider the types of distillation problems normally encountered in the course of an engineering education. They are usually concerned with binary systems requiring easily defined separations. Later, when the student is introduced to multicomponent separations, the systems are almost invariably mixtures of light hydrocarbons, mainly because there is a reasonable amount of fairly accurate equilibrium data in the literature with which to demonstrate the calculation procedures. The distillation of petroleum fractions is discussed hurriedly, if at all, and the concept of the relationship between reflux and trays to separation capability is seldom considered in detail. Where the student or even experienced engineer is accustomed to think of distillation in terms of percent recovery, allowable impurity levels and the like, the refinery engineer uses terms such as initial and final boiling points, boiling range, gap specifications and so on. This would lead the casual observer to the conclusion that different procedures would be required to design petroleum fractionators as opposed to conventional chemical systems. To a certain extent this is true, but there is also a great area of common ground. An understanding of this will afford a much greater appreciation of the design and operation of petroleum fractionators.

It is the purpose of this chapter to consider distillate fractionator design in the terminology of the refinery and simultaneously to show the similarity to classical distillation design principles. A secondary objective is to consider the representation of these separations by a McCabe-Thiele graphical analysis so that the operation and key component profile throughout the tower may be more easily visualized.

This chapter assumes that the reader has a reasonable knowledge of the properties of various petroleum fractions and products and of the methods available for estimating physical properties and physical equilibria. Where this is not the case, the first five of the cited references are recommended. The other references discuss design fundamentals considered in this chapter.

Introduction

A light ends fractionation unit can be found in several locations within the refinery processing scheme. Many refineries will have several units in this service, their multiplicity being dictated by being keyed into various types of processing units. They are *always* found in the following places.

1. In conjunction with a crude unit since the naphtha must be debutanized to permit its being stored at atmospheric pressure.
2. In conjunction with a catalytic cracking unit since the raw catalytic naphtha is rich in propylene and butylenes which are used as alkylation unit feedstocks. As in No. 1 above, a secondary consideration is the reduction of the naphtha vapor pressure.
3. In conjunction with a coking unit for the same reasons as in No. 2.
4. In conjunction with a hydrocracker which ordinarily yields considerable volumes of light ends in addition to the liquid product being produced.

It is possible to design a light ends unit to process feeds from several plants, and this is often done, particularly in the case of building a grass roots refinery or in the case of a major expansion involving several new units. In taking this approach, one key rule must be observed.

> Saturated and unsaturated hydrocarbons are never fed to the same gas plant, the reason being that the saturated light ends are most often sold as fuel while the unsaturated light ends are used as feedstocks to other units.

Thus, one might design a saturates gas plant to process the total naphtha from a crude unit plus light ends from all other sources of saturated hydrocarbons within the complex. This could include hydrotreaters, hydrocrackers, isomerizers and catalytic reformers. The unsaturates gas plant would be designed to process total light ends from all sources of unsaturated hydrocarbons. This could include catalytic cracking units, cokers or any thermal decomposition unit. In this latter case, naphtha fractionators would probably be provided for each naphtha since, of those mentioned, only catalytically cracked naphtha is of suitable quality for sale without further processing or treating.

In the following sections are listed various hydrocarbons and naphthas found in the refinery and the purposes for which they are used.

Saturated Light Ends—Products and Specifications

Products and End Uses

1. Methane:
   - fuel gas
   - hydrogen unit feed gas
2. Ethane:
   - fuel gas
   - ethylene unit feedstock
3. Propane:
   - fuel gas
   - liquid fuel (LPG)
   - ethylene unit feedstock
   - alkylation unit feedstock (when mixed with propylene)
4. Isobutane:
   - fuel gas
   - liquid fuel (LPG)
   - petrochemical feedstock (pyrolysis to propylene and isobutylene)
   - alkylation unit feedstock (when mixed with butylenes)
   - gasoline blending (vapor pressure and octane improvement)
5. N-butane:
   - fuel gas
   - liquid fuel
   - ethylene unit feedstock
   - gasoline blending
6. Isopentane:
   - gasoline blending (octane improvement; clear research octane number = 92.3)
7. N-pentane:
   - gasoline component (low octane; clear RON = 61.7)
   - isomerization unit feedstock (conversion to isopentane)
   - catalytic reformer feedstock (part of stabilized naphtha)
8. N-hexane:
   - gasoline component (low octane; clear RON = 24.8)
   - isomerization unit feedstock (conversion to C_6 isomers)

TABLE 5.1

Maximum Impurity Compositions		
Component	Liq. Vol. %	Mole %
Ethane	4.1	4.4
Propane	93.4	93.5
Isobutane	2.5	2.1
TOTAL	100.0	100.0

catalytic reformer feedstock (part of stabilized naphtha)

9. C_6 Isomers (often referred to as "isohexanes")
   high octane gasoline blending:
      2.2 dimethylbutane (RON clear = 91.8)
      2,3 dimethylbutane (RON clear = 101.7)
   Low octane gasoline blending or isomerization unit feedstock:
      2 methylpentane (RON clear = 73.4)
      3 methylpentane (RON clear = 74.5)

Product Specifications

The bulk of the propane and butanes which are produced as refinery liquids are sold as fuel in the form of liquefied petroleum gas (LPG). Detailed specifications for these materials have been published by the Natural Gas Processors' Association (NGPA). These specifications cover many aspects of product quality and composition, many of which are outside the control or influence of the distillation portion of the total manufacturing process. For both propane and butane, two variables are defined which are composition dependent and which are used as bases for separations in towers yielding these products.

Propane. The vapor pressure as measured at 100 degrees F shall not exceed 200 psig, and the isobutane content shall not exceed 2.5 liquid volume percent. These correlate to the maximum impurity compositions given in Table 5.1.

Butane. The vapor pressure as measured at 100 degrees F shall not exceed 70 psig, and the isopentane content shall not exceed 1.95 liquid volume percent. Table 5.2 shows the maximum impurity level compositions for the indicated ratios of iso- to normal butane.

Ethane. Ethane has become very valuable for use as an ethylene plant feedstock. Ascrizzi (6) has summarized the desirable compositions and properties for this application.

Other Components. The specifications for other possible light ends products are usually set either by purity requirements for downstream processing units or by economic design considerations when being used as gasoline blendstock.

Unsaturated Light Ends—Products and Specifications

Products and End Uses

1. Ethylene:
   petrochemical feedstock
2. Propylene:
   petrochemical feedstock
   alkylation unit feedstock (when mixed with propane)
3. Butylenes:
   petrochemical feedstock
   alkylation unit feedstock (when mixed with isobutane)
4. Butadiene:
   petrochemical feedstock

Product Specifications

Since these products are most often used for feeds to petrochemical units, their purity specifications are often quite high—99 percent and even higher being quite common. These specifications are usually established by the processing requirements of the receiving unit and, thus, can seldom be treated as design variables in the fractionation step of feed preparation.

Straight-run Naphthas

Motor Gasoline

Usually 100 to 400 degrees F, ASTM containing sufficient front end volatility—usually as butane—for suitable ignition but not so much as to produce vapor locking. Ten to 12 psia is a good target vapor pressure for the finished product.

Catalytic Reformer Feed for Motor Gasoline Production

Usually 100 to 375 or 385 degrees F ASTM containing basically isopentane through the C_{10}'s. Care must be taken to reject all C_{11}'s from this stream since the conversion of C_{10} paraffins and naphthenes to C_{10} aromatics increases the ASTM end point by 15 to 25 degrees F. This will vary somewhat with the nature of the naphtha under consideration.

TABLE 5.2

Composition Variables of NGPA Grade Butanes					
	Case				
	A – C_4 = 100% iC_4		B – C_4 = 100% nC_4		C – General Mixture of iC_4 and nC_4
Component	Liq. Vol. %	Mole %	Liq. Vol. %	Mole %	Liq. Vol. %
Propane	9.80	11.4	21.80	24.3	21.85 – .138 X
Isobutane	88.25	86.9	–	–	X
N-butane	–	–	76.25	74.1	76.2 – .862 X
Isopentane	1.95	1.7	1.95	1.6	1.95
Total	100.00	100.0	100.00	100.0	100.0

Catalytic Reformer Feed for Petrochemical (BTX) Production

Precursors for benzene boil over a TBP range of 156 to 177 degrees F. Precursors for toluene boil over a TBP range of 189 to 245 degrees F. Precursors for the xylenes boil over a TBP range of 245 to 270 degrees F. This would indicate an ASTM boiling range of approximately 150 to 250 degrees F. Nelson (OGJ, Feb. 3, 1969, 104-105) discusses this in detail.

Naphtha for Cracking to Petrochemicals

As pointed out by Nelson (OGJ, March 10, 1969, 76-77), there are many criteria, economic as well as technical, which influence the compositions of desirable cracking feedstocks. The main properties desirable in a cracking feedstock are high naphthene, low aromatic, low isoparaffin straight-run materials and low sulfur content. Desirable molecular weights usually range from 80 to 120.

Fractionation Specifications for Naphthas

General composition requirements for the fractionation of light ends or discrete components have been discussed in previous sections of this chapter. Setting fractionation requirements for these towers generally reduces to solving a material balance problem to yield products meeting predetermined and, usually, rather rigid composition criteria. Thus, the material covered here will be limited to the types of separations encountered in the fractionation of continuous boiling petroleum liquids. The procedures discussed consider the separation to be a split between two discrete components, their recoveries being determined by economic considerations. This is accomplished by breaking up the feedstock as described by a TBP distillation into pseudocomponents which have a relatively narrow boiling range characteristic. As an illustration, consider a mixture of C_6, C_7 and C_8 paraffinic and naphthenic hydrocarbons. All C_6 paraffins have pure component boiling points ranging from 122 to 156 degrees F; all C_7 paraffins boil between 175 and 209 degrees F; and all C_8 paraffins boil between 211 and 258 degrees F. Cyclohexane boils at 177 degrees F; all C_7 naphthenes boil between 190 and 218 degrees F; and all C_8 naphthenes boil between 221 and 269 degrees F. Thus, for this illustration, that portion of the TBP curve boiling over the range 122 to 177 degrees F is C_6; the portion boiling between 175 to 218 degrees F is C_7; and the portion boiling between 211 to 269 degrees F is C_8. Note that the final point of the lighter component is not exactly the same as the initial point of the heavier component. This anomaly is overcome by setting the TBP cut point between pseudocomponents at the midpoint between the two temperatures. No matter how the required fractionation is specified, it can readily be calculated by applying classical light ends methods to the assumed pseudocomponents.

These separations can be specified in one of four ways. Although they may appear to be radically different, they result in the same general analysis after the fractions are broken up into their respective pseudocomponents. These four different types of fractionation specifications are discussed as follows.

Light Product ASTM End Point and ASTM (5-95) Gap Specification

This is the type of specification normally encountered in refinery work. The split is solved by trial-and-error to find the proper combination of TBP end point of the lighter product and TBP overlap at the cut point which will satisfy the ASTM temperature requirements. The end result is a separation based on light key recovery on the distillate and heavy key recovery in the bottoms.

Boiling Ranges of Products

In this type of specification, ASTM boiling ranges are stated for the product streams. By converting these to TBP curves, one then estimates the distribution of the key pseudocomponents and, thence, the separation. Note that the ASTM (5-95) Gap cannot be specified in this case, but rather results from plotting the ASTM curves.

Split between Components

In this type of specification, the separation is stated to be a complete split between two components. This is obviously not possible. The design material balance is set up by cutting the feedstock so that the total volume of the distillate is the same as the total volume of the desired light ends recovery. The amount of heavy ends in the distillate is established by setting the distillate TBP end point at the final TBP temperature of the light component boiling range. This results in the minimum amount of heavy key in the distillate consistent with yielding only the total light ends volume overhead. The TBP initial point of the bottoms is established by assuming that the total TBP overlap at the cut point is distributed equally between the products. Once the split is set up, it reduces to a light and heavy key recovery specification.

Recovery of Keys

In this type of specification, the required recovery of light and heavy keys in the distillate and bottoms, respectively, is stated directly. Alternately, one may specify recovery of the major key in a product stream, the other being defined as an allowable impurity concentration in that stream.

Types of Light Ends Plants

As was discussed in the previous section, light ends plants are designed to process *only* either saturated or unsaturated hydrocarbons. Only in the rarest of cases would these be mixed to form feed for a single unit. However, within the framework of establishing design criteria for a gas plant, there are yet other factors which must be considered in developing the process.

The first of these obviously is the desired/required number of product streams to be yielded, their compositions or purities, specified yields or recoveries of any particular component(s), the physical state–either vapor or liquid–in which the products must be delivered to other systems or to storage, the availability and condition of utility streams and, quite often, economic criteria which must be observed. These points suffice to define the required number of towers and approximate temperature and pressure levels at key points throughout the system. The arrangement of the towers will be determined by the distribution of products within the overall feed stream. This subject will be considered in detail in the next section of this chapter.

The second and much more subtle factor considers the relative quantities which are present of components which are lighter than the lightest product. This situation occurs in gas plants for hydrocrackers, catalytic cracking units, cokers and natural gas processing units. This latter type is not found within the refinery and is not considered here in any degree of detail but is based on the same design principles now being discussed. In these types of units, one finds that a large percentage of the total feed consists of components lighter than the lightest product. In the case of the first three units mentioned above, this will include hydrogen as well as hydrocarbons. In the natural gas processing plant, it may include nitrogen, carbon dioxide, hydrogen sulfide and occasionally helium. In the case of an FCCU main fractionator design, ethane and lighter may make up as much as 10 to 12 mole percent of the gross feed to the main tower which can equate to as much as 20 mole percent of the feed to the gas plant. It is obvious that processing such a feedstock in a conventional deethanizer would result in the following.

1. Yielding ethane and lighter as a vapor distillate at a maximum pressure level and using a water-cooled condenser would result in high losses overhead of the C_3's in the feed.
2. Reducing C_3 loss would require operating a refrigerated condenser which, in turn, requires a costly feed drying system.

These two undesirabilities can be avoided by using a reboiled absorber as the first tower in such a gas plant followed by the product fractionators. A sharp separation is made by using an absorption oil to recover the heavy key in the bottoms and reboiled stripping to reject the light key to the overhead vapor. The design of this type of tower will be discussed in the final section of this chapter. A typical gas plant utilizing a reboiled absorber for an FCCU fractionation section, coker or hydrocracker is shown by Figure 5.9. A typical configuration for a natural gas processing plant utilizing a reboiled absorber is shown by Figure 5.8.

The various configurations of refinery light ends units in addition to other valuable qualitative information can be found in Bozeman's (8) excellent paper.

Process Design Considerations

This section will discuss the basic principles involved in designing a single light ends tower and/or a complete processing unit yielding several products. Detailed design procedures for different types of towers will be presented in later sections of this chapter. Some of the material presented may seem overly basic, but, quite often, elementary principles are either neglected or treated hastily, resulting in substandard designs or in work which must be redone due to the assumption of erroneous bases.

Basic Design Considerations

Assuming that the chemical engineering function of the unit has been determined, i.e., the separation(s) to be made have been established, the first order of business is to determine how this can best and most economically be accomplished. Since all distillation processes involve the addition of and removal of heat, the proper selection of process temperature levels becomes the first and most important job of the designer. It is obviously desirable to remove heat from the condenser by using either cooling water or air-fin coolers. This is true even in the case of difficult separations where one might be inclined to operate a refrigerated condenser to achieve a lower operating temperature profile and, thus, a lower relative volatility which will decrease tray requirements. Refrigeration is rarely found in refinery gas plants but is quite common in modern natural gas processing plants and in petrochemical plants such as ethylene units. Having defined the separation and the distillate condensing temperature, the bottoms temperature now follows. One must then choose the proper heating medium.

In the refinery, there are usually three sources of heat—steam, high temperature oil which must be cooled before being stored and direct fired heaters. Steam is the first choice for bottoms product temperatures up to approximately 350 degrees F. Above this level, either hot oil or a direct fired reboiler will be required. The following guidelines should be observed in selecting operating utilities.

1. When installing new facilities in an existing refinery, determine whether or not there is any excess capacity in the existing utility systems and how much is available for the design in question. Utilize this spare capacity wherever possible.
2. If the cooling water system has insufficient spare capacity for the design, air-cooled condensers and product coolers are usually cheaper than additions to the refinery cooling water system.
3. In recent years, the trend has been toward maximizing air-cooler usage and minimizing cooling water usage when designing grass-roots facilities. This would seem indicated in revamps or major expansions as well.
4. Minimizing steam consumption by using fired heater reboilers will minimize steam plant and boiler feedwater preparation facilities. An alternate approach is to use a hot oil loop where one furnace fires enough heat for all reboilers on the service.
5. The excess heat contained in hot process streams should be used with caution. In the varying operating conditions and product slates encountered in the average refinery, the availability of the stream being considered as a heat source may not always be adequate. In crude unit design, especially, the author prefers using hot products for crude feed preheat and for auxiliary reboiling only when the hot stream contains at least 50 percent more heat than required by the controlling reboiler duty.

These are not the only points to be considered but are indicative of the type of thinking which must be done prior to commencing actual design work.

Arrangement of Processing Steps—Unit Arrangement

This discussion applies only to situations where two or more towers in sequence are being considered. Single towers are designed on the basis of the available feed, but in any multiple product distillation train, there are a number of ways in which the towers can be arranged. Consider, for example, a three-component system consisting of propane, isobutane and n-butane which must be separated into relatively pure components. Two routes can be used. The first tower can be a depropanizer yielding a propane product distillate followed by a C_4 splitter (deisobutanizer). Alternately, the first tower can be a deisobutanizer yielding a n-butane bottoms product followed by a depropanizer.

In the first scheme, the depropanizer would be designed to fractionate between propane and isobutane. Since this is a relatively easy split, the tower could contain a moderate number of trays. The C_4 splitter would contain a large number of trays since the separation is relatively difficult.

In the second scheme, the first tower would contain a relatively large number of trays since it would be designed for the isobutane-n-butane separation. The second tower, the depropanizer, would contain a moderate number of trays.

The operating conditions and relative equipment sizes for the two are shown in Figure 5.3. By inspection, it is obvious that Method 1 is the better process design because:

1. Equipment costs will be lower since the large tower will be designed for a lower pressure.
2. Operating costs will be lower since the deisobutanizer

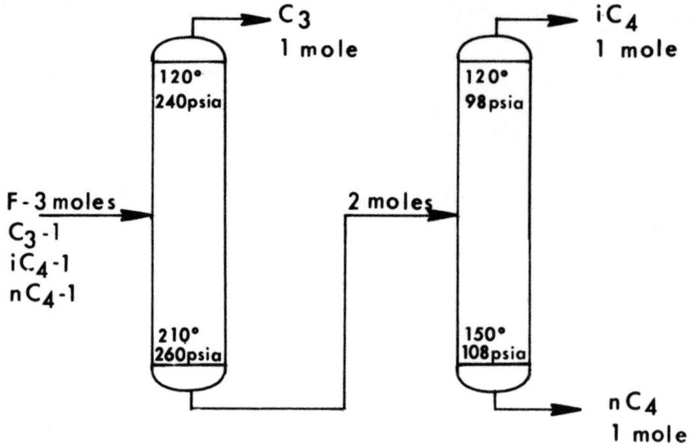

RELATIVE NO. OF TRAYS	1	2
RELATIVE TOWER DIAM.	1	2.61
TOWER DESIGN PRESSURE, psig	300	120

METHOD I

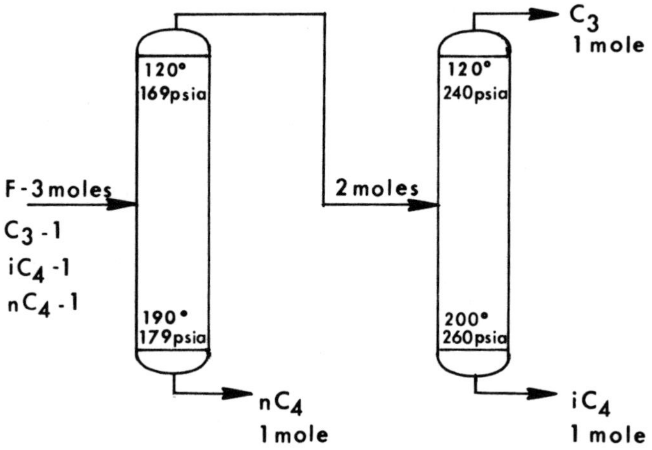

RELATIVE NO. OF TRAYS	2	1
RELATIVE TOWER DIAM.	3.02	1
TOWER DESIGN PRESSURE, psig	210	300

METHOD 2

Figure 5.3. Process design alternates for production of propane, isobutane and n-butane.

condenser load will be approximately half that of Method 2. This will be reflected in a lower reboiler heat duty for Method 1.

This type of analysis should be applied in the arrangement of any multitower fractionation train. Harbert (7) addressed himself to this problem and arrived at three rules-of-thumb for arriving at the most economical arrangement of towers in any given fractionation plant.

1. "Try and achieve a 50-50 split of the feed into distillate and bottoms." If the quantities of the two product streams are made as nearly equal as possible, consistent with the required material balance, the heat needs of the two halves of the column also approach equality, and the heat degraded across the tower is used to the maximum, once in each half of the tower. This leads to minimum heat needs for the separation.
2. "Minimize the charge to a tower where a difficult separation is required." The feed to a tower having high reflux requirements and, thus, high heat requirements should be minimized. This will result in minimizing equipment sizes and utility consumptions.
3. "Total heat inputs to all the towers of a system provide a good measure of the process effectiveness." Consider a given fractionation system operating at conditions of minimum reflux (and infinite trays). The tower arrangement which results in minimum thermal cost—considering both the cooling of reflux and reboil heating—will usually define the optimum arrangement of towers. This also applies to systems requiring refrigerated reflux where one must consider compression as well as heat transfer costs.

In the example of Figure 5.3 which selects Method 1 as the better design, Rule 2 was used as the principal criterion. By inspection, heat loads become minimized, and Rule 3 confirms that Method 1 is the better choice. However, the real impact of the above rules can best be seen in the design of an overall fractionation unit.

Consider Figure 5.4, which is a flowsheet for a complex refinery saturates gas plant based on 100,000 BPSD crude oil feed to the refinery. The material balance for this operation is given by Table 5.3. The arrangement of processing steps is typical of many refineries, particularly of those which have grown by expansion and revamp.

Now consider Figure 5.5 and Table 5.4, the flowsheet and material balance, respectively, for a unit processing the same feedstock into the same products but arranged in accordance with Harbert's rules. By inspection, one can quickly see the savings in both capital and operating costs which can be achieved by an intelligent approach to process layout.

Feed Preheat

A further section of Harbert's paper considers the questions of preheating the feed to a tower and how much feed preheat is desirable. He points out that the feeling among designers is that tower feed should be preheated. He states, "It is usually thought that the feed should be hot and one rule that the writer has heard is that the [mole?] percent of the feed vaporized should equal the [mole?] percent of the overhead product." He goes further to point out that the heat requirements of any given tower are determined by the separation being made rather than by the heat added to the process by the feed.

To further explore this question, a parametric study was made by this author to determine the effect of the thermal condition of the feed upon the overall process design of a tower and the heat quantities involved. This work was based on analyzing three binary hydrocarbon systems so that the calculations could be easily treated by hand. However, the relationships developed also apply to multicomponent systems.

Heat Required by Distillation Processes

This subject has been treated thoroughly throughout the literature, but, for sake of clarity, it is represented here.

Basic Heat Effect. In reference to Figure 5.6,

$$F = D + W$$

D leaves the tower as a dew point vapor, and W leaves the tower as bubble point liquid. The heat effect incurred by this process is expressed as

$$\Delta H_1 = Q_{DV} + Q_{WL} - Q_F$$

Note that the heat removed in condensing D to a bubble point liquid plus the amount of heat contained in D as a bubble point liquid is equal to the heat content of D as a vapor.

Heat Effect Due to Minimum Reflux. Calculations can be made to determine the minimum reflux ratio required to affect the desired separation for a given thermal condition of the feed. This heat effect takes into account the heat removed by the condenser, Q_{CM}, in condensing the distillate plus the minimum amount of reflux required by the process and is expressed as

$$Q_{CM} = (D + R_M) \times (H - h)$$

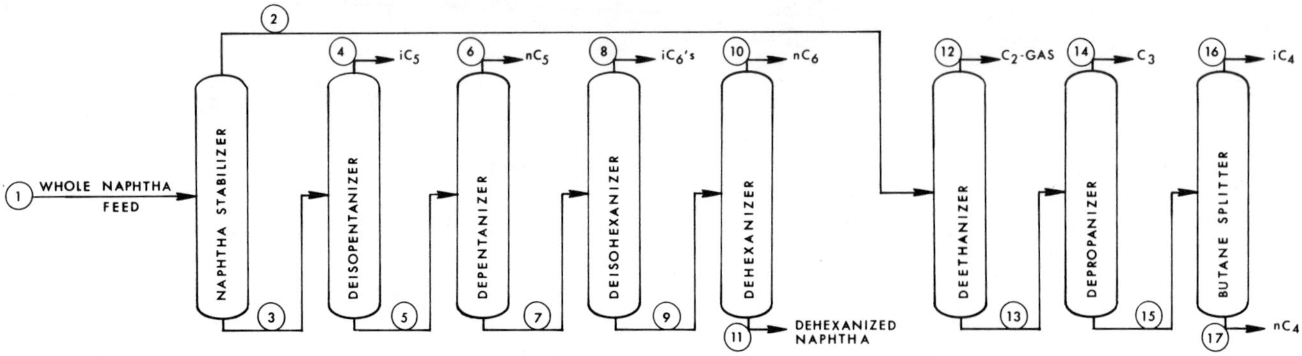

Figure 5.4. Typical flow scheme, refinery saturates gas plant.

TABLE 5.3

Material Balance Summary for Refinery Saturates Gas Plant—Typical Flow Scheme (Figure 5.4)

Stream No.	1	2	3	4	5	6	7	8	9	10	11	12	13	14	15	16	17
Component																	
C_2	35.55	35.55										34.66	.89	.89			
C_3	124.22	124.22										3.03	121.19	118.08	3.11	3.11	
iC_4	59.58	58.98	.60	.6									58.98	1.47	57.51	54.63	2.88
nC_4	119.53	116.54	2.99	2.99									116.54	1.17	115.37	5.77	109.60
iC_5	59.96	1.50	58.46	49.20	9.26	8.68	.58	.58					1.50		1.50		1.50
nC_5	62.69	.63	62.06	9.08	52.98	51.43	1.55	1.55					.63		.63		.63
cyc. C_5	32.51		32.51		32.51	.81	31.70	28.53	3.17	3.17							
22 DMB	6.63		6.63		6.63	.07	6.56	5.25	1.31	1.31							
23 DMB	10.81		10.81		10.81	.05	10.76	7.53	3.23	3.23							
2 MEP	62.72		62.72		62.72		62.72	18.82	43.90	43.90							
3 MEP	39.61		39.61		39.61		39.61	7.92	31.69	31.69							
nC_6	82.40		82.40		82.40		82.40	8.03	74.37	72.31	2.06						
C_6H_6	24.52		24.52		24.52		24.52	2.39	22.13	21.52	.61						
C_7 +	1179.95		1179.95		1179.95		1179.95		1179.95	7.37	1172.58						
Total, moles/hour	1900.68	337.42	1563.26	61.87	1501.39	61.04	1440.35	80.60	1359.75	184.50	1175.25	37.69	299.73	121.61	178.12	63.51	114.61

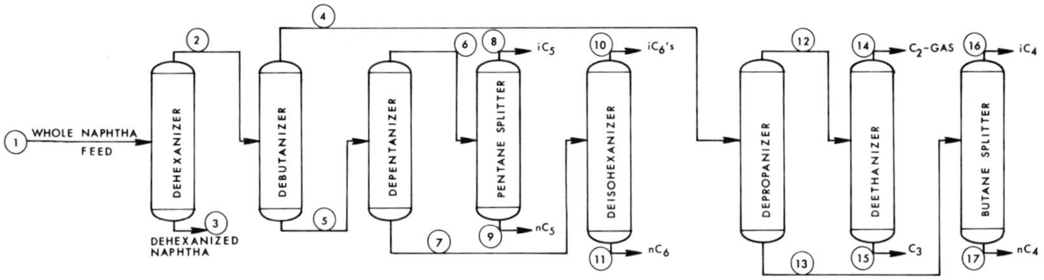

Figure 5.5. Alternate flow scheme, refinery saturates gas plant.

TABLE 5.4

Material Balance Summary for Refinery Saturates Gas Plant—Alternate Flow Scheme (Figure 5.5)

Stream No.	1	2	3	4	5	6	7	8	9	10	11	12	13	14	15	16	17
Component																	
C_2	35.55	35.55		35.55								35.55		34.66	.89		
C_3	124.22	124.22		124.22								121.11	3.11	3.03	118.08	3.11	
iC_4	59.58	59.58		58.98	.60	.60		.60				1.47	57.51		1.47	54.63	2.88
nC_4	119.53	119.53		116.54	2.99	2.99		2.99				1.17	115.37		1.17	5.77	109.60
ic_5	59.96	59.96		1.50	58.46	57.88	.58	49.20	8.68	.58			1.50				1.50
nC_5	62.69	62.69		.63	62.06	60.51	1.55	9.08	51.43	1.55			.63				.63
cyc. C_5	32.51	32.51			32.51	.81	31.70		.81	28.53	3.17						
22 DMB	6.63	6.63			6.63	.07	6.56		.07	5.25	1.31						
23 DMB	10.81	10.81			10.81	.05	10.76		.05	7.53	3.23						
2 MEP	62.72	62.72			62.72		62.72			18.82	43.90						
3 MEP	39.61	39.61			39.61		39.61			7.92	31.69						
nC_6	82.40	80.34	2.06		80.34		80.34			8.03	72.31						
C_6H_6	24.52	23.91	.61		23.91		23.91			2.39	21.52						
C_7 +	1179.95	7.37	1172.58		7.37		7.37				7.37						
Total, moles/hour	1900.68	725.43	1175.25	337.42	388.01	122.91	265.10	61.87	61.04	80.60	184.50	159.30	178.12	37.69	121.61	63.51	114.61

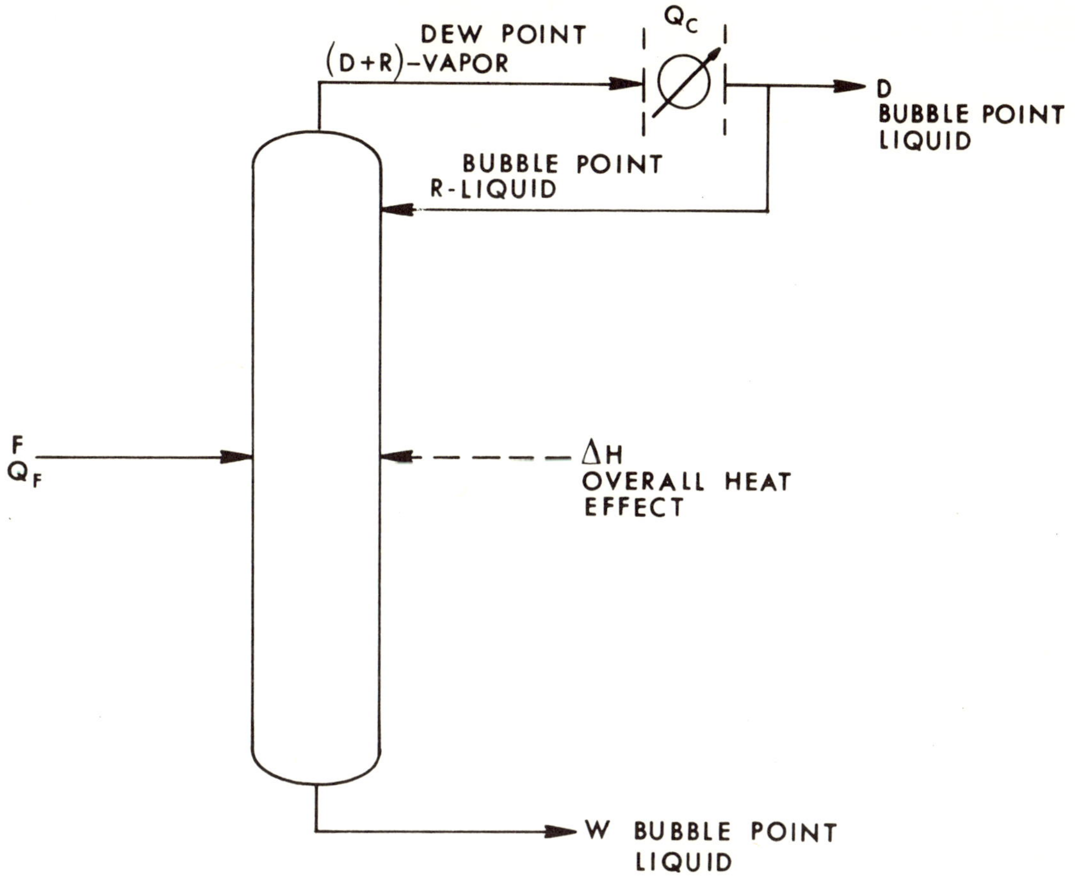

Figure 5.6. Heat requirements for a typical one feed-two product distillation.

where R_M = minimum reflux calculated for the system.

This is the minimum heat requirement for the process and is expressed as

$$\Delta H_{RM} = (Q_{DL} + Q_{WL} - Q_F) + Q_{CM}$$

Heat Effect Due to Actual Reflux. The additional reflux above the minimum requirement incurs an incremental duty in condensing this extra vapor. Thus,

$$Q_{CA} = (D + R) \times (H - h)$$

where R = actual reflux for the system. The total heat requirement for the operating system is then expressed as

$$\Delta H_{RA} = (Q_{DL} + Q_{WL} - Q_F) + Q_{CA}$$

This heat is added to the system by one or more reboilers.

Basis of Study

The study considered the following range of variables.

Input Data

1. Three binary systems: difficult (propylene-propane) @ α = 1.17; average (propane–n-butane) @ α = 2.375; easy (propane–n-pentane) @ α = 5.22.
2. Three different feed compositions: 0.25, 0.50 and 0.75 light component mole fraction.
3. Two different feed thermal conditions: dew-point vapor and bubble-point liquid.

TABLE 5.5

Raw Data—Fractionation Study

System																		
X_F, mol frac. Lt. Comp.	.75		.50		.25		.75		.50		.25		.75		.50		.25	
Thermal Cond. of Feed	SV	SL	SV	SL	SV	SL	SV	SL	SV	SL	SV	SL	SV	SL	SV	SL	SV	SL
N_M, Min Trays	49.573—All Cases						9.000—All Cases						4.81—All Cases					
R_{DMIN}	8.33	7.63	13.20	11.56	27.82	21.79	1.05	.633	2.29	1.14	5.90	3.71	.568	.106	1.47	.319	4.16	.90
R_{DDSG}	9.79	8.97	15.51	13.59	32.69	25.60	1.23	.744	2.69	1.34	6.93	4.36	.667	.125	1.73	.375	4.89	1.05
NTP	102.2	102.2	100.8	101.2	99.7	100.1	21.2	23.1	19.8	21.7	19.4	19.8	16.6	16.5	11.9	14.3	11.1	12.5
NT@EOC = .7	144.6	144.6	142.5	143.1	141.1	141.6	28.9	31.6	26.9	29.6	26.3	26.9	22.3	22.1	15.5	19.0	14.4	16.4
Q_C, M^2BTU/hr 100 mol F	4.563	4.215	4.591	4.055	4.487	3.543	1.018	.794	1.105	.701	1.139	.769	.770	.519	.828	.418	.856	.299
Q_R, M^2BTU/hr 100 mol F	3.970	4.222	3.988	4.062	3.879	3.550	.331	.831	.405	.750	.431	.805	.051	.677	.085	.640	.126	.502
Top Tray V,																		
V, mol/hr	821	758	826	729	807	637	170	133	184	117	190	128	127	86	136	69	141	49
L, mol/hr	745	682	776	679	783	613	94	57	134	67	166	105	51	9.5	86	19	117	25
Bottom Tray,																		
V, mol/hr	721	758	726	729	707	637	70	133	84	117	90	128	27	86	36	69	41	49
L, mol/hr	745	782	776	779	783	713	94	157	134	167	166	205	51	109.5	86	119	117	125
Average Rel. Vol.	1.17 – All Cases						2.375 – All Cases						5.220 – All Cases					

Assumptions

1. Constant compositions of distillate and bottoms at 0.98 and 0.02 mole fraction of light component, respectively.
2. Operating reflux ratio = minimum reflux ratio x 1.175.
3. Constant molal overflow, i.e., the internal material balance was not determined by heat balance calculations. For these systems which are very close to ideal, the error inherent in this assumption is small.

Values Calculated for Each System. The following quantities were calculated for each feed composition at each feed thermal condition.

1. Minimum reflux ratio.
2. Minimum trays required at total reflux.
3. Condenser duty per 100 moles feed.
4. Reboiler duty per 100 moles feed.
5. Number of theoretical trays at finite reflux.
6. Number of actual trays assuming 70 percent overall tray efficiency.
7. Vapor-liquid traffic at all key trays.

Calculated Results and Conclusions

Table 5.5 illustrates the raw data and Table 5.6 the comparative data obtained from the calculations. For the case of dew-point vapor feed, it is noted that this mode of operation:

TABLE 5.6

Comparative Data—Fractionation Study

System: $C_3 = -C_3$

$(X_F)C_3 =$	.75	.50	.25
$(R_{SV}/R_{SL})_D$	1.0920	1.1420	1.2770
$(N_{SV}/N_{SL})_{TP}$	1.0000	.9958	.9965
$(A_{SV}/A_{SL})_{V.\ L.}$	1.0830	1.1330	1.2670
$(Q_{SV}/Q_{SL})_R$	.9403	.9818	1.0930

System: C_3, nC_4

$(X_F)_{C_3}$	.75	.50	.25
$(R_{SV}/R_{SL})_D$	1.6590	2.0090	1.5900
$(N_{SV}/N_{SL})_{TP}$	.9180	.9120	.9800
$(A_{SV}/A_{SL})_{V.\ L.}$	1.2780	1.5370	1.4840
$(Q_{SV}/Q_{SL})_R$	.3983	.5400	.5354

System: C_3, nC_5

$(X_F)_{C_3}$	.75	.50	.25
$(R_{SV}/R_{SL})_D$	5.3580	4.6080	4.6220
$(N_{SV}/N_{SL})_{TP}$	1.0060	.8320	.8880
$(A_{SV}/A_{SL})_{V.\ L.}$	1.4770	1.9710	2.8780
$(Q_{SV}/Q_{SL})_R$	.0753	.1328	.2510

1. Requires a greater reflux ratio and condenser duty. The disparity becomes greater as the separation becomes easier.
2. Requires fewer operating trays. The disparity becomes greater as the separation becomes easier.
3. Requires a lower reboiler duty in every case but one. The disparity becomes greater as the separation becomes easier.
4. Requires greater column area with the load tray being in the rectifying section. Thus, if swaging is not practical, the only tower design economy can be effected by reducing tray spacing in the stripping section.

The converse of all the above is true for the case of bubble-point liquid feed.

Based on the above facts, the following conclusions are reached for design cases.

1. The practice of preheating column feed to save reboiler duty requires a greater condenser duty and greater tower area. Thus, the designer must consider utility costs for heating and cooling as well as equipment costs. Generally speaking, more condensing surface is required per unit of heat transferred than is heating surface. This is especially true in the case of fired reboilers, although alloy requirements for tubes may sometimes offset the apparent savings differential.
2. In economic climates where cooling is expensive due to water shortage or to high ambient air temperatures, e.g., the Middle East, feed preheat by feed-bottoms exchange will almost always result in higher capital and operating costs.
3. Feed preheat should *never* be practiced in columns having refrigerated condensers. Rather, maximum feasible precooling of the feed is indicated, especially if refrigeration recovery can be effected.
4. In difficult separations, minimizing the heat content of the feed will materially reduce reflux requirements and tower area with no significant increase in tray requirements.

For revamps of or additions to existing plants, the problems are not so easily defined as in a design situation. In many cases, systems are first limited by factors other than the tower, such as pumps, condensers, reboilers, external heat exchangers, etc., then the tower itself. Other factors may come into play such as plot area availability or the limitation to an expensive piece of equipment such as a fired heater. Utilities system limitations may force the designer into a course of action which ordinarily might not be indicated. The following general guidelines are recommended for revamps/expansions.

1. If a reboiler is limiting and the tower is not, capacity can be increased up to the limit of the tower by increasing feed preheat and additional condensing surface.
2. If the tower is limiting, additional capacity may be obtained by reducing the heat content of the feed, thereby decreasing the heat load on the condenser at the expense of increased reboiler duty. A detailed analysis of the internal vapor-liquid traffic is required to optimize this case, and access to a computer can greatly simplify the study.

In many cases, particularly in older plants, towers are operating at conditions far different than those for which they were designed. A study of such towers will often reveal that the owner is getting far more out of the tower than calculations can predict. Unfortunately, test runs are rarely possible in such cases due to various system limitations. If a test run can be made, this is the surest way to define capacity, but, more often than not, the engineer will have to rely on calculations plus the experience of the owner, both being tempered by judgment.

Example Design Calculations

In this section are presented procedures for calculating every type of distillation process to be found in any refinery gas plant. These are

1. Simple light ends fractionator
2. Whole naphtha stabilizer
3. Whole naphtha splitter
4. Complex fractionator for whole naphtha
5. Reboiled absorber.

Each procedure will be presented as a narrative explaining in detail the required steps. These, together with the others presented in this work, describe design methods for all types of refinery fractionation processes. This work does not consider the absorber since this is very rarely found in the refinery. In addition, the literature is well stocked with absorber design procedures; only new or, at least, previously unpublished work is presented here.

These methods will emphasize "short cut" types of calculations which enable the designer to obtain a close approximation of the answer prior to using more sophisticated techniques if such are indicated to refine the design. There are available at this time many highly developed computer programs for solving distillation problems, but many of these require as input data estimates of temperature, profiles of vapor and liquid traffic, number of trays, etc. The methods presented here can also be used as generators of preliminary data for computer input. An accurate first estimate can often result in significant savings in computer time—which is money—and in actual time which also is money. A second benefit is that supporting hand calculations enable one to analyze computer output much more rapidly and to spot discrepancies much more easily.

Calculation of Simple Discrete-Component Fractionators

This type of column is found in the light ends section of the refinery gas plant and processes only identifiable components. It is also the common type of fractionator found in the distillation section of natural gas liquids recovery facilities or, for that matter, in most chemical plants where the process objective is to make a separation between two components. The design procedure for calculating this type of tower is outlined in the following discussion.

Process Design Basis

As in any engineering design work, the first question to be resolved is "What is it that we are trying to accomplish?" or, in the case of distillation design, "What kind of separation is required by the process?" The designer will generally be guided by criteria such as recovery requirements for certain components, allowable impurity levels in streams, economic values assigned to the various recovered streams and the like, all being overshadowed by whatever economic factors may exist which are unique to the project under consideration.

At the beginning of the design, the engineer must first set the material balance for the tower. This is most often done on the basis of making a separation between two components whose volatilities, i.e., boiling point or vapor pressure characteristics relative to each other, make them adjacent in the listing of the components in the feed. These two components are called *key components* and are defined as follows.

> "The Light Key Component is the heaviest component whose percent recovery is greater in the distillate than in the bottoms. Conversely, the Heavy Key Component is the lightest component whose percent recovery is greater in the bottoms than in the distillate.

Obviously, key component recovery can never be 100 percent. If it were, another component would be the key component.

In a typical case, the tower could be a depropanizer making a separation between propane and isobutane. Purity specifications on the butane product from the base of the tower allow a relatively large concentration of propane. Isobutane level in the recovered propane would be estimated by what separation would occur in a deethanizer which would further fractionate the depropanizer distillate into an ethane-rich fuel gas overhead and a propane product bottoms. The preceding knowledge is not required in order to perform actual calculations but is included here to show the relationships that come into play in setting up a design.

Having defined the key components and their required recoveries, one can now establish the preliminary material balance for the design. Ordinarily, one can assume that all components lighter than the light key will appear completely recovered in the distillate and that all components heavier than the heavy key will be completely recovered with the bottoms. In the case where the separation is to be between components which have isomers, such as between propane and the butanes, isobutane being the heavy key, a portion of the non-key isomer will follow along with the key. In the preliminary estimate of the material balance, the designer should guess the non-key component distribution because the presence of these components may materially affect the temperatures and pressures within the system. These assumptions will be checked later and then adjusted for the final calculations.

Occasionally, a situation will arise where the key components are not adjacent when ranked according to volatility. In this case, components lying between the keys will appear in substantial quantities in both product streams and, for this reason, are called *distributed components*. In setting up the preliminary material balance, the designer must estimate the quantitative presence of these components in the products. This preliminary assumption will later be checked by the techniques described in a later portion of this procedure.

Operating Temperatures and Pressures

The operating conditions at key points in the system are established in the following sequence.

Reflux Drum Conditions. By definition of the process requirements, it is usually known that the distillate must be either a vapor or a liquid. From this knowledge, the reflux drum conditions are set as a function of the available cooling media.

If cooling water is to be used in the condenser, set the temperature in the reflux drum at the maximum allowable temperature of the outlet cooling water. For example, cooling water is available at 85 degrees F, and its return temperature shall not exceed 120 degrees F. Therefore, set the reflux drum temperature at 120 degrees F.

In systems which are inherently high pressure, the operating pressure level may be significantly reduced by taking an approach of 15 degrees F at the cold end of the condenser and a 15 degree F rise on the cooling water. For the conditions stated previously, the reflux drum would operate at 100 degrees F, and the cooling water temperature range would be 85 to 100 degrees F.

If the condenser is to be cooled by air, set the temperature in the reflux drum no closer than 40 degrees F to the available summer dry bulb temperature. For example, air is available at 90 degrees F dry bulb. This will result in a minimum reflux drum temperature of 130 degrees F.

In the case where the condenser duties are large and where there is a genuine need to reduce tower pressure by cooling the overhead to the minimum feasible temperature, the condenser function may be satisfied by a combination of air and water cooling provided that the temperature of the vapor leaving the top tray is sufficiently high. For example, vapor could be cooled-condensed from 200 to 130 degrees F, using 90 degrees F air, and from 130 to 100 degrees F, using water from 85 to 100 degrees F. This technique minimizes cooling water consumption, at the same time maintaining reasonable equipment sizes. Having established the temperature in the reflux drum, the pressure is calculated by bubble-pointing in the case of liquids or dew-pointing in the case of vapor distillates.

Top Tray Temperature. For the case of a liquid distillate, the composition of the vapor leaving the top tray is the same as that of the distillate product. Thus, the top tray temperature is determined by making a dew point calculation on the distillate composition after taking a pressure increase of 5 to 10 psi above that in the reflux drum. For the case of a vapor distillate, the condenser is the top separation stage, and its temperature is already known. Later, after the reflux requirements have been determined, the top tray temperature is calculated by adding the reflux to the vapor distillate and dew-pointing the resultant mixture after taking the necessary pressure increase. For vapor distillate systems, this is necessary only to specify the condenser and is not required for the distillation analysis.

Bottoms Temperature. The bottoms temperature is its bubble point after taking a pressure increase of 5 to 10 psi across the trays. In the case of relatively low pressure operation, the actual pressure which can be obtained and which becomes known only after the equipment design is complete may necessitate a reevaluation of the assumed pressure profile. This is seldom critical in towers operating above 100 psia. A design criterion that must be satisfied at this point is that the base temperature should not approach the pseudocritical temperature of the bottoms product by any closer than 25 degrees F. If this situation arises, two alternatives are normally open. First, consider producing the distillate as a vapor using the conventional condenser coolant. This will result in operating the reflux drum at the same temperature as before but at the dew-point pressure.

If, however, the production of a liquid distillate is mandatory, set the base temperature at 25 degrees F less than the pseudocritical temperature of the bottoms and establish the pressure levels in the system. After calculating the temperature required in the reflux drum, select a refrigerant temperature level to give a cold end temperature difference at the condenser of 15 to 25 degrees F.

A third alternate is to yield the overhead product as two phases, their composition being determined by making a flash calculation at the appropriate temperature dictated by the available cooling media at this maximum operating pressure level.

A second criterion that must be observed is that the pressure in the reflux drum must not fall below 5 psig. This may necessitate operating the reflux drum at a temperature higher than that normally attainable with the available cooling media. This practice is limited to systems whose distillate product is hexanes or heavier. i.e., whose atmospheric pressure boiling points exceed 120 degrees F.

Checking the Preliminary Material Balance

The original assumptions as to the distribution of nonkey components are now checked by employing Hengstebeck's (16) technique which postulates that all components in a distillation are distributed according to the relationship,

$$\log (D/W) = f(\log \alpha)$$

This is used in the following sequence.

1. Calculate the arithmetic average temperature and pressure of the tower based on the vapor from the top stage and the liquid from the bottom stage.
2. At this condition, tabulate the vapor-liquid equilibrium coefficient, the K value, for each component in the system. Then, calculate their volatilities relative to the heavy key component.
3. For the key components only, calculate D/W, which is their ratio of distribution between distillate and bottoms.
4. On log-log paper, plot the points, (α, D/W), for the two keys and draw a straight line between them. The ratio of distribution for all components in the system will lie on this line. At any given value of α, the corresponding value of D/W for that component is found which, for the purpose of this illustration, will be designated as z.

$$D/W = z, \text{ which reduces to } D = zW = z(F - D)$$

$$D = F\,[Z/(1 + z)]$$

5. If revisions are required to the assumed material balance, the temperatures and pressures within the system must be checked for the new balance and revised where necessary. Unless a serious error is made in the original assumptions, more than one iteration will seldom be required. This procedure is particularly useful in systems having distributed components.

Thermal Condition of the Feed

The thermal condition of the feed to the tower is generally not treated as a design variable except in a revamp situation where one might attempt to control vapor or liquid loadings in the tower by controlling the feed flash. Since feed comes to the tower system from another system, its thermal condition upon entering the environment of the tower is a function of its temperature and pressure level into which the feed enters. The thermal content of the feed can be altered by heat exchange against other streams, by heating or by cooling utilizing plant utilities. The implications to design of the thermal condition of the feed have been treated in detail earlier in this chapter. For the purpose of final design involving any feed heat exchange, one can calculate the feed bubble point and dew point at tower pressure and then a flash at some intermediate temperature for the purpose of drawing a heating curve.

Average Relative Volatility

The average relative volatility for the light key-heavy key pseudobinary system is determined by calculating these values at the conditions of dew-point vapor leaving the top stage and bubble-point liquid leaving the bottom stage and then using the lesser of geometric and arithmetic mean values. The geometric mean will usually be the smaller value and the one to use.

The average relative volatility will be used in all calculations involving the determination of tray and reflux requirements.

Minimum Number of Stages at Total Reflux

The minimum number of separation stages required to attain the specified separation at conditions of total reflux is calculated by Fenske's (11) equation.

N_M = [log (light key in distillate x heavy key in bottoms) / (light key in bottoms x heavy key in distillate)] ÷ log α.
= log (LKD x HKW) / (LKW x HKD) ÷ log α

An alternate method is that of Winn (12), which can be used as a check. The two methods agree within reasonable tolerances, but Winn's method is a little more time consuming to use.

Minimum and Operating Reflux Ratios

The literature is full of methods and procedures for calculating the minimum reflux required to effect a specified separation. Most are complex and require a good deal of time, quite often involving trial-and-error. Of all those known to this author, Underwood's (10) procedure is the simplest to use for manual calculations.

The method recommended here is an analytical solution for the reflux-to-distillate ratio which has been found to check very closely the values obtained from other more sophisticated calculation procedures. It has been of particular value in setting up and analyzing distillation problems on the computer. It involves a three-step procedure correlating key component vapor-liquid equilibrium at the feed point and the purity of the reflux liquid.

A stepwise illustration for the case of bubble-point feed is as follows. In this discussion, note that the x and y values refer to the light key mole fractions in a pseudobinary system consisting only of the keys. These concentration values are

x_R = mole fraction in reflux
x_F = mole fraction in total feed
x_{FV} = mole fraction in feed point equilibrium liquid
y_{FV} = mole fraction in feed point equilibrium vapor

1. Calculate the light key concentration in the vapor phase which is in equilibrium with the bubble point feed. This quantity is

$$y_{FV} = (\alpha\, x_F) / [1 + (\alpha - 1)\, x_F]$$

2. Calculate the following quantity

$$m = (x_R - y_{FV}) / (x_R - x_F)$$

3. Calculate the minimum reflux-to-distillate ratio as

$R_{DM} = m/(1 - m)$, moles keys in reflux/moles keys in distillate.

A similar treatment for dew point vapor feed yields these sequential equations.

$$X_{FV} = X_F / [\alpha - (\alpha - 1)\, X_F]$$
$$m = (X_R - X_F) / (X_R - X_{FV})$$
$$R_{DM} = m/(1 - m)$$

For feeds which are partially flashed, the value of R_{DM} can be closely approximated by interpolating between the values for saturated liquid and saturated vapor feeds in direct proportion to the percentage of total keys vaporized in the feed flash.

Also note carefully that the reflux ratio is expressed as *moles of keys in reflux/moles of keys in distillate*. In the case of a liquid distillate, the ratio is a straightforward total moles of reflux/total moles of distillate. In the case of vapor distillate, however, the total moles of reflux will be greater than the value which is calculated as the ratio of the total moles of distillate since the distillate vapor is richer in keys than the reflux liquid.

Over the years, studies of many light hydrocarbon fractionators have established that the economic optimum reflux ratio ranges from 1.15 to 1.25 times the minimum. For separations where the relative volatility is low or where there is a degree of uncertainty as to its accuracy, a ratio of 1.2 to 1.3 times the minimum is often employed.

Estimate of Number of Trays at Finite Reflux

Gilliland's (13) correlation is recommended for estimating the number of theoretical separation stages required by the design. An alternate method is that of Mason (14). Gilliland is quite simple and is used as follows.

1. Calculate $(R_D - R_{DM}) / (R_D + 1)$.
2. From Figure 5.7, obtain the value of $(N - N_M) / (N + 1) = K_G$.
3. Calculate the number of 100 percent efficient separation stages as

$$N = (N_M + K_G) / (1 - K_G).$$

We are now faced with the problem of equating theoretical to real stages. Two rules apply.

1. A partial condenser is counted as one 100 percent efficient separation stage.
2. A reboiler may count as one 100 percent efficient separation stage only if it processes the total liquid from the bottom tray on a once-through basis, yielding boilup vapor and bottoms liquid. If this criterion is not met, take no tray credit for the reboiler. Observing rules 1 and 2 to reevaluate the number of stages required as tower trays, one now applies a tray efficiency which is nothing more than a judgment factor to obtain the number of actual trays to be specified for the tower.

McCabe-Thiele Analysis

The feed tray location is obtained by making an analysis of the key component pseudobinary system similar to the familiar McCabe-Thiele procedure (9). Hengstebeck has also shown ways to treat multicomponent systems by this technique. The calculational sequence is as follows.

1. Calculate and plot the $x - y$ equilibrium envelope as

$$y = \alpha x / [1 + (\alpha - 1) x]$$

2. Draw in the feed line at the appropriate degree of vaporization. Note that the line is vertical for bubble point feed and horizontal for dew point feed. For partially vaporized feed, a flash calculation is necessary to determine the light key concentration in the vapor and liquid phases as y_{FV} and x_{FV}, respectively; these are not the same y_{FV} and x_{FV} values discussed earlier in this procedure. Having determined these values, the feed line is drawn in from point (x_{FV}, y_{FV}) to point x_F on the 45 degree line.
3. Draw in the rectification section operating line from point x_R on the 45 degree line to point

$$[0, X_R / (R_D + 1)].$$

4. Draw in the stripping section operating line from point x_W on the 45 degree line to the intersection of the rectification section operating line and the feed line.
5. By the familiar stairstep technique, obtain the number of trays in each section and apply a suitable tray efficiency, remembering to take credit for the condenser and/or reboiler if this is indicated.

Overall Heat Balance

The shortcut design is completed by calculating the heat quantities in the feed, distillate, bottoms and the condenser duty. The reboiler duty is then calculated by difference as

$$Q_R = (Q_C + Q_D + Q_W) - Q_F$$

Calculation of Fractionation of Discrete Components and Hydrocarbon Mixtures Defined by a Continuous Boiling Range

This column is quite often the first processing step in the refinery gas plant and, in some cases, is the only naphtha fractionation step used. It is also found in most conversion units, i.e., reformers, hydrotreaters, etc., for remov-

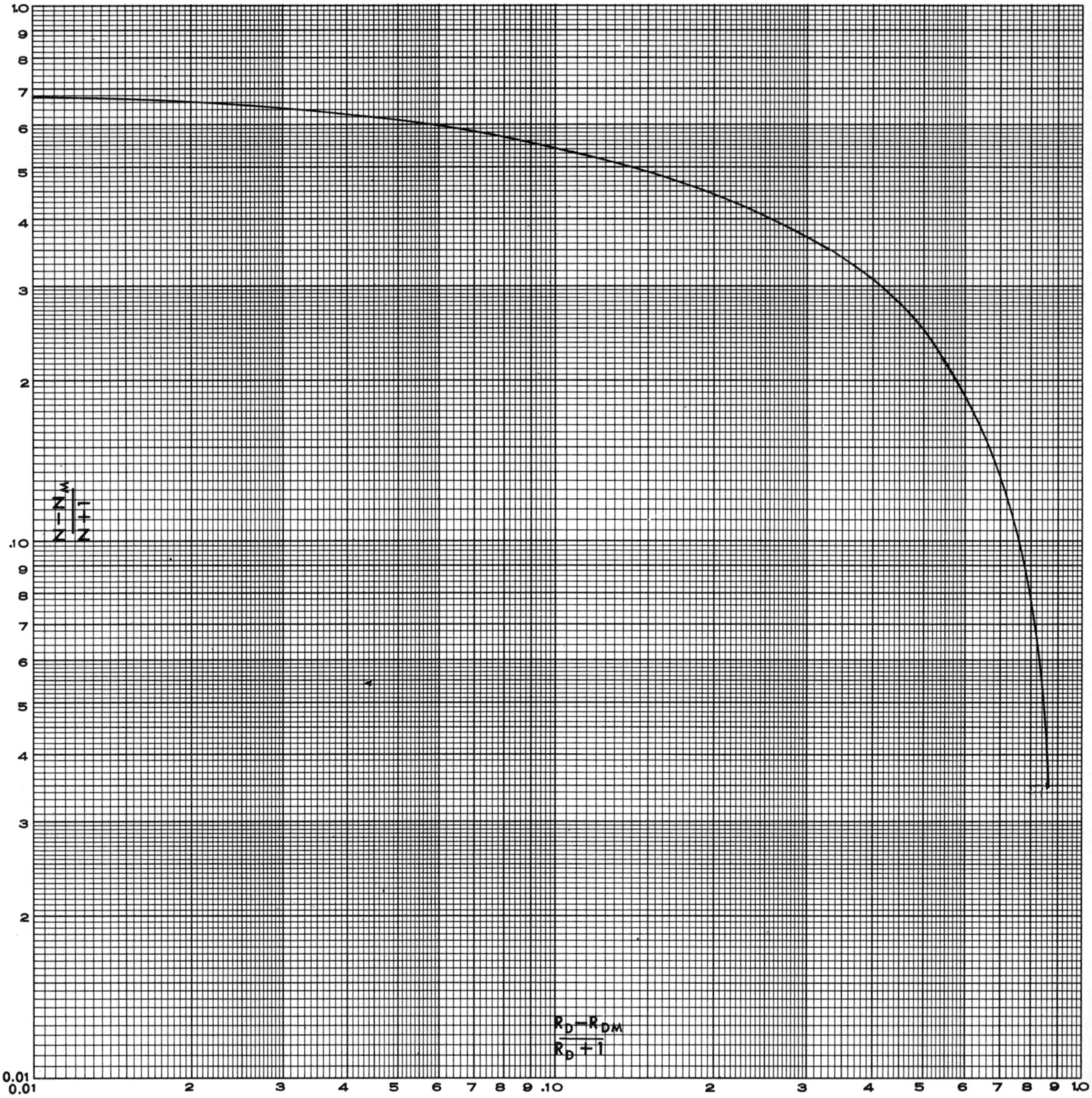

Figure 5.7. Gilliland's correlation, theoretical to real trays (used with permission of the American Chemical Society).

ing light ends formed in the process from the heavier liquid product. It may also be found in the oil production areas in use as a crude oil stabilizer. In this latter service, the column removes the light ends as a distillate liquid so that the bottoms product will have a vapor pressure less than atmospheric when stored at ambient temperature. The process objective is to make a sharp separation between discrete components, but, in this case as opposed to the previous example, the major portion of the tower feed consists of hydrocarbon liquid having defined physical properties rather than a componential description. The design procedure for calculating this type of tower is outlined in the following discussion.

Process Design Basis

This question is a little easier to resolve here than in the case of discrete-component fractionators since the only real criterion for product quality is vapor pressure of the bottoms product. Thus, there is no economic driving force to remove any more light components than necessary to meet vapor pressure specifications, especially since butane is blended back into the gasoline to ensure a certain vapor pressure. It is also a fact that, in most economic climates, butane commands a higher price as gasoline than as LPG. Given a feedrate and its characteristics, it is first necessary to establish the approximate componential structure before one can define the tower material balance.

Feedstock Definition

The properties of the feedstock can be quoted in one of two ways. The usual way is by an ASTM or TBP distillation and the liquid specific gravity. Alternately, it may be described by a light ends analysis up through the pentanes with the remainder being described as "C_6 and heavier"; a specific gravity is also required here. If experimental data on the feedstock is lacking, one can determine the above-mentioned properties from an assay of the parent crude oil.

Now, from the overall distillation, either TBP or ASTM, and the specific gravity of the feed, previously discussed techniques (3), (4) and (5) are employed to determine the following basic information.

1. If given the TBP distillation, convert it to an ASTM or vice versa if given the ASTM distillation.
2. Atmospheric EFV distillation.
3. Characterization factor, K.
4. Average molecular weight.
5. Pseudocritical temperature and pressure.
6. Phase diagram plot.

Pure Components in Feedstock

If at all possible, the concentration of light ends in the feed, ethane through the pentanes, should be taken directly from the parent crude oil assay. In the next section which describes the whole naphtha splitter, methods will be presented for estimating the componential composition for the whole feed, including light ends. Having determined the light ends makeup of the feed, tabulate the properties and flow rates of materials to be fed to the tower.

Design Separation

As pointed out earlier, separation criteria are somewhat less stringent than in pure-component light ends fractionation. Establishing the design separation for a naphtha debutanizer, for example, might reduce to a study of how much butane can be left in the bottoms without exceeding vapor pressure specifications on the gasoline. If, however, the naphtha debutanizer bottoms were to be used as catalytic reformer charge stock, then one might make as sharp a separation as possible between the butanes and pentanes in order to minimize feed to the reformer. This is the type of thinking involved in setting up one of these towers.

In a typical design, the separation might be defined as a recovery percentage of n-butane in the distillate to define the light key and an allowable level of isopentane in the distillate to define the heavy key. From this point on, the design separation is defined by the following sequential steps.

1. Knowing the required light key recovery and assuming that all components lighter than the light key will appear entirely in the distillate, tabulate the light keys and lighter in the distillate.
2. Add in the allowable heavy key to the distillate.
3. Calculate the volume percentage of the feed which is to be yielded as distillate. This will establish the TBP cut volume and temperature point on the TBP plot of the feedstock.
4. For the purpose of establishing the distillation curves of the tower bottoms product, one can assume with only a very small possibility of error that the TBP initial point of the bottoms is equal to the normal boiling point of the light key and that the TBP end point of the distillate is equal to the normal boiling point of the heavy key. From these two points, draw in the TBP curve of the bottoms product.
5. Convert the bottoms TBP curve to ASTM and EFV curves. Calculate the properties of the bottoms and plot its phase diagram.
6. Calculate and tabulate the properties of the feed, distillate and bottoms streams.

From this point on, much of the material has been presented and discussed in the previous section. Thus, further discussion will be limited to new concepts.

Operating Temperatures and Pressures

1. Set the operating temperature in the reflux drum.
2. Calculate the reflux drum pressure by making either a bubble-point or dew-point calculation on the distillate as indicated.
3. In the case of a liquid distillate, calculate the top tray vapor temperature.
4. Take a reasonable pressure increase across the tower and determine the temperature of the bottoms liquid product from the phase diagram. This condition is bubble point liquid at the total pressure in the base of the tower.

5. Calculate the average relative volatility for the pseudobinary system made up of the key components.
6. Calculate the minimum number of trays required for the separation at conditions of total reflux.
7. Calculate the minimum reflux ratio required for the separation as indicated by the thermal condition of the feed. Note that the flash curve for the feed can be obtained from the feed phase diagram at the feed tray total pressure.
8. Set the operating reflux-to-distillate ratio.
9. Estimate the number of theoretical stages required for the separation at the design reflux ratio. Apply a suitable tray efficiency and determine the actual number of trays to be provided.
10. Make a McCabe-Thiele analysis for the design separation to check the number of trays required and to locate the feed tray.
11. Calculate the overall heat balance for the system.

Calculation of Fractionation between Two Defined Hydrocarbon Mixtures

This column in the refinery gas plant is used to fractionate further the debutanized naphtha from the bottom of the naphtha debutanizer into fractions having the required characteristics. An alternate use is as the first tower in the refinery gas plant where it yields a full-range light naphtha distillate and a heavy naphtha bottoms. In this latter case, the distillate would require further stabilization in another tower. The reasoning behind this configuration would be to optimize the gas plant design in accordance with Harbert's guidelines. The design procedure of calculating this type of tower is outlined in the following discussion.

Process Design Basis

In this case, the properties of both streams are described in terms of their distillation characteristics even though the feedstock will contain significant quantities of identifiable light ends. This is because the design separation will be a split between light and heavy fractions rather than between the light ends and a defined mixture as in the previous section.

The properties of the products and the fractionation requirements are usually defined in terms of their ASTM end points, ASTM boiling ranges, ASTM (5-95%) gaps and the like.

Feedstock Definition

The feedstock is described and defined in accordance with the criteria previously outlined. As was the case there, it is necessary to determine the same properties in order to define the feedstock for further calculations.

Componential Definition of Feedstock

The composition of the feedstock is estimated by dividing the continuous TBP curve into volumetric portions whose boiling ranges describe the pure component boiling points of all materials present having the same number of carbon atoms. This does not mean that the C_6 cut of any given stream will contain a mixture of C_6 paraffins, naphthenes, olefins and aromatics. Rather, the origin of the feedstock will give a clue to its composition.

A virgin naphtha from the crude still would be expected to contain mostly paraffins and naphthenes with the possibility of some aromatics, depending upon the nature of the parent crude.

Another common naphtha, catalytic reformate, would contain mostly isoparaffins and aromatics. The characterization factor of the mixture will usually categorize its composition. A value of 10 means a high degree of aromaticity while a value of 12 means high paraffinicity. Naphthenes have K values in the range of 11, but this could also describe a mixture of aromatics and paraffins. Thus, the K value and a knowledge of the origin of the feed will aid in estimating the categories of the components present.

Once this has been resolved, the whole feed is broken up into pseudocomponents by the following technique.

1. For all components *possibly present* which have the same number of carbon atoms, tabulate their pure component boiling points as the boiling ranges of all the isomers of a given category. As an example, the lightest C_6 paraffin boils at 122 degrees F and the heaviest at 156 degrees F. The comparable temperatures for the C_6 olefins are 106 degrees F and 164 degrees F. Thus, a mixture of only C_6 paraffins would be characterized by a boiling range of 122 to 156 degrees F. A mixture of C_6 paraffins and olefins would have a boiling range of 106 to 164 degrees F, which also includes the 122 to 156 degree F segment.
2. After the presence of all possible types of components has been established, select the applicable boiling ranges. For heavy fractions, there is usually a temperature overlap between light fraction end point and heavy fraction initial point. The TBP cut point between the two is arbitrarily taken as the mid-point between the two. For light fractions, there may be a temperature gap between the initial point of the first pseudocomponent, which has a boiling range, and the heaviest discrete component, which has a pure component boiling point. In this case, the TBP cut point between the two is taken as the midpoint. When this has been done, the liquid volume percent of all com-

ponents present by the number of carbon atoms will have been established.

3. For each component in the feed, tabulate the volumetric quantities present. Using the properties of n-paraffins, calculate the weight and mole quantities present in the feed to the tower. There will almost always be a discrepancy between the gravity and molecular weight calculated here and those values previously derived. This is because of the error inherent in the assumption that the feed contains only normal paraffins when it has been clearly established that it is a mixture of several types. Unfortunately, making this assumption is the only way to obtain weight and gravity data on the feedstock pseudocomponents other than resulting to a laboratory investigation. This latter approach is seldom economically justifiable. The error is of little real consequence since, as will be seen later, the equilibrium data used in the fractionation calculations will be based on vapor pressure data derived from the bulk properties of the pseudocomponents. In order to avoid confusion, adjust the molal flow rates of each component present so that the total will equal the actual number present as calculated earlier.

Design Separation

Depending upon the way in which the fractionation requirements are defined, estimate the design separation and tabulate the distribution of components between the product streams and note the recovery of light and heavy key components.

Product Properties

From the TBP curves used in deriving the design split and assuming that the characterization factor of the products is the same as that of the feed, derive the ASTM and EFV distillations, the API gravity, the average molecular weight, the pseudocritical properties and the phase diagrams. The molecular weight of the bottoms should be calculated as the difference between feed and distillate.

Operating Conditions

Using the distillate phase diagram and the design criteria covering the condenser cooling media, set the operating conditions in the reflux drum as the distillate bubble point pressure at the appropriate temperature. Taking a suitable pressure increase to the top tray, set the temperature at the distillate dew point. Taking a suitable pressure increase to the base of the tower, set the temperature of the bottoms products as its bubble point of the indicated pressure.

Note that the limiting criteria discussed earlier for bottoms temperature and reflux drum pressure must also be observed in these types of towers.

1. Calculate the average relative volatility for the pseudo-binary system made up of the key components.
2. Calculate the minimum number of trays required for the separation at conditions of total reflux.
3. Calculate the minimum reflux ratio required for the separation as indicated by the thermal condition of the feed.
4. Set the operating reflux ratio.
5. Estimate the number of theoretical stages required at the operating reflux ratio. Applying a suitable tray efficiency, set the number of actual trays to be provided.
6. Make a McCabe-Thiele analysis to check the number of theoretical stages and to locate the feed tray.
7. Calculate the overall heat balance for the system.

Calculation of Fractionators Producing Three Defined Hydrocarbon Mixture Products

This type of column is common in catalytic reformers, hydrotreaters and desulfurizers but can be used wherever it is desirable to produce an overhead distillate, either liquid or vapor, sidedraw liquid distillate and a bottoms product. It can also be used as the first step in naphtha fractionation.

The procedures are similar in many respects to those of earlier sections, so this discussion will be devoted to points of dissimilarity or to principles not previously covered. The first three steps are calculated exactly as in the previous section.

1. Process design basis
2. Analysis of feedstream
3. Breakdown of feed into pseudocomponents

Design Separation

This part of the calculation will be handled as before, except that a split must be set between the distillate and the sidecut and between the sidecut and the bottoms.

Product Properties

For each product, the required product properties are calculated and a phase diagram is drawn.

Operating Conditions

Draw a process flowsheet and set up a material balance for the tower showing all pseudocomponents. As before, the phase diagrams are used to determine operating temperatures and pressures.

Thermal Condition of Feed

Calculate the degree of feed flash at its temperature and pressure.

At this point, all the information has been developed to enable the fractionation calculations to begin. Referring back to the previous example, the procedure was to calculate the operating reflux ratio from which the condenser duty was established. The reboiler duty was then calculated by making an overall system heat balance. In a two-distillate product column, however, the required reflux ratios must exist at both the top tray and the sidedraw tray. Another feature of this type of tower is that the lower section where the feed enters at some intermediate tray is a two-product tower in its own right. This section yields the sidecut as a top tray liquid by the same mechanism as in a crude tower. The vapor passing this tray contains all the distillate components plus the vapor necessary to satisfy the system heat balance. The latter portion contains the components of the sidecut. Thus, the upper section is a rectifying section whose feed must be established from calculations around the lower section.

This method first calculates reflux and tray requirements for the base section in order to establish upper section feed. The upper section is then analyzed for reflux requirements which are compared to reflux availability, a function of the system heat balance. The detailed procedure is as follows.

Lower Section Design

1. For the two keys of the bottoms-sidecut fractionation, calculate the average relative volatility using pseudocomponent vapor pressures at the sidecut draw tray and bottoms temperatures.
2. For the required fractionation, calculate the minimum reflux ratio and set the operating reflux at 1.25 times the minimum.
3. Calculate the minimum number of trays required by Fenske's equation.
4. From Gilliland's correlation and an appropriate efficiency, estimate the actual tray requirements.

Upper Section Design

1. For the upper section, draw a process flowsheet and set up a component material balance. Note that the bottom tray of the upper section is the top tray of the lower section and that the upper section bottoms product is the sidecut. The material balance is based on the distillate product, the sidecut and the reflux from the sidecut draw tray. This has the same composition and is at the same temperature as the sidecut. The composition of the vapor flowing upward to the sidecut draw tray is established by material balance and is the upper section feed. Calculate the light key mole fraction in the combined key pseudobinary system.
2. For the two keys of the distillate-sidecut fractionation, calculate the average relative volatility using pseudocomponent vapor pressures at the top tray and sidedraw tray temperatures.
3. Calculate the minimum reflux for the required separation and set the operating reflux ratio at 1.25 times the minimum. For the operating reflux, calculate the required condenser duty.
4. Calculate the minimum number of trays at total reflux.
5. Estimate the number of theoretical and actual trays.
6. Calculate the dew point temperature of the vapor feed to the upper section using K data derived from pseudocomponent vapor pressures.
7. Recalculate the condenser duty based on an upper section overall heat balance and compare with the value calculated in Step 3. If this value of Q_c is greater than the previous value, the reflux available to the top section is greater than that required for that separation, and the sidecut-bottoms separation is controlling. Thus, the design is satisfactory as is. If the results indicate that satisfying the lower section reflux requirements will not provide sufficient upper section reflux, the design will have to be reworked in accordance with the following section. Assuming for the moment that the design is satisfactory, proceed to Step 8.
8. Calculate the heat input to the system required from the reboiler by making an overall system heat balance.
9. Using the revised operating reflux ratio, reestimate the actual tray requirements.
10. Make a McCabe-Thiele analysis of both sections in order to establish overall tray requirements and the locations of the feed and draw trays.

Redesign Procedure (If Required)

In Step 7 above, it may be established that upper section reflux requirements will control the design. If so, the following procedure is to be followed.

1. Set the upper section reflux ratio at 1.25 times the minimum and calculate the condenser duty.
2. By a trial-and-error procedure, solve the upper section overall heat balance by assuming a value for reflux from the sidecut draw tray. Calculate the upper section feed by material balance and then find its dew point. When the heat balance calculation meshes, compare the light key mole fraction of the combined key pseudo-

binary feed to the upper section with that value used in previous Step 1. There will seldom be a significant difference here. If there is, however, recalculate the reflux requirements for this new separation and repeat the heat balance. One iteration will usually be sufficient to firm up the design for the upper section.
3. Reestimate the number of trays in the lower section using the new reflux ratio.
4. Make a McCabe-Thiele analysis of both sections to set the overall tray requirements and key tray locations.

Reboiled Absorber

The following is a discussion of reboiled absorbers in various services and a suggested manual procedure for calculating the absorption section. The method is simple, relatively rapid and has the same degree of accuracy as a computer analysis. The method can also be used for estimating lean oil rates to the top of the tower as a starting point for computer calculations. Reboiled absorbers are notoriously difficult machine calculations unless relatively loose convergence limitations are specified. Thus, an accurate first estimate of lean oil rates could result in significant savings of computer time.

Introduction

The reboiled absorber is an item of mass transfer equipment which is often found in the gas processing sections of all types of process plants, although it is more common in the natural gas processing and petroleum industries. It is used to make a separation between heavy materials which are recoverable at moderate temperatures and pressures and to reject all lighter components. By using absorption-stripping rather than distillation, the separation can be effected at less severe conditions. For example, a typical column makes a separation between ethane and propane at 350 psia by using 40 degree F lean oil. If this same separation were accomplished by distillation, a temperature level of −30 degrees F would be required. Or, at a top temperature of 40 degrees F, a column operating pressure of approximately 1,200 psia would be necessary. Thus, the advantages of this type of separation are easily seen.

The descriptive term *reboiled absorber* is somewhat misleading in describing the function, although it is the name most commonly used. Another descriptive name and the one most often used in the hydrocarbon processing industries is *rich oil demethanizer* where high ethane recovery is attained or *rich oil deethanizer* when propane recovery is practiced; in either case, the abbreviation *ROD* is often used. The most accurately descriptive name but, unfortunately, the one least used now is *stripper-reabsorber*.

The function of the reboiled absorber is to make a separation between two components recovering the desired amount of the heavier as part of the bottoms liquid and rejecting the lighter of the two to the overhead gas stream. As in distillation, these two components are called the heavy key and the light key, respectively. Note that all components present which are heavier than the heavy key will be recovered in the bottoms liquid. Liquid to the top tray of the tower is provided by a cool, relatively heavy oil. This is used to wash back heavy components in the upflowing gas stream by the mechanism of absorption rather than by condensing and refluxing a part of the column overhead stream as in a distillation operation. The lean oil is usually either a special oil heavier than any of the components in the tower feed, or it will correspond to the composition of the heaviest product stream made from the feed. The only requirement, however, is that the lean oil must be heavier than the light key.

Feed enters the column as a vapor-liquid mixture containing all the components of the system. The degree of feed vaporization which is desirable plays an important role in the economics of the design and will be discussed later in greater detail. Optimization of the feed flash balances the maximization of light key in the vapor while minimizing its heavy key content. The light key and lighter components are stripped out of the liquid falling from the feed tray, and this internal vapor, together with the feed flash vapor, flows upward from the feed tray to the absorption section where the heavy key and heavier components are absorbed in the lean oil. The estimation of the composition and volume of the stripping section exit vapor is the key to the design and is the principal subject of this section.

In the hydrocarbon processing industries, the reboiled absorber is most often found in natural gasoline plants and refinery gas plants. In certain circumstances, one might be found in the recovery section of a pyrolysis plant. Figures 5.8 and 5.9 illustrate the role of the reboiled absorber in two different types of units, both of which are designed to recover propane as the lightest salable liquid. In the gasoline plant, the raw gas from the driers is cooled and then chilled ahead of the high pressure absorber. Hydrocarbon condensate is knocked out before the gas enters the absorber, and this liquid is part of the ROD feed. The gas portion enters the absorber where cold lean oil absorbs a predetermined amount of the propane. For the purposes of this discussion, it is assumed that essentially all the butanes and heavier components are also absorbed as well as some of the ethane and lighter components. The removal of these lighter materials while minimizing propane loss to the residue gas is the function of a rich oil deethanizer. In modern plants, the absorber operates in the range of −30 degrees F. Refrigeration is recovered by exchanging the warm lean oil against the cold rich oil. This heats the rich oil to approximately 80

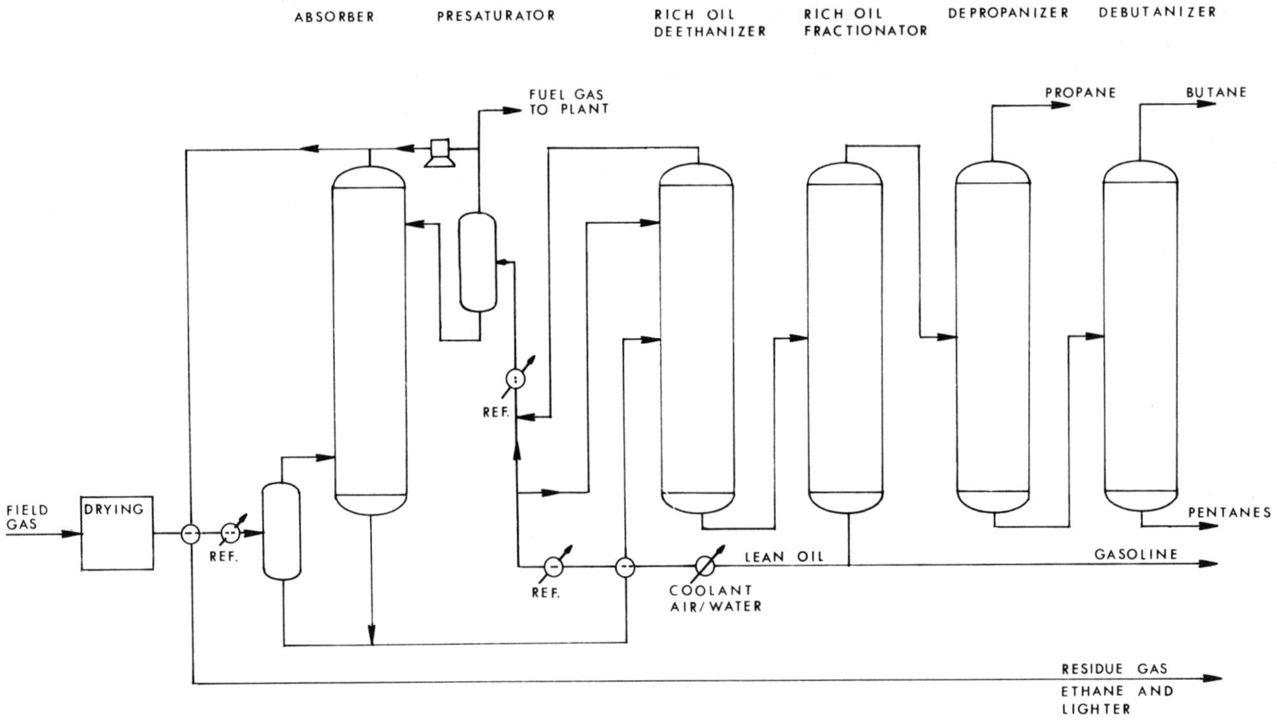

Figure 5.8. Natural gas processing plant for recovering propane and heavier. Note: condensers and reboilers not shown.

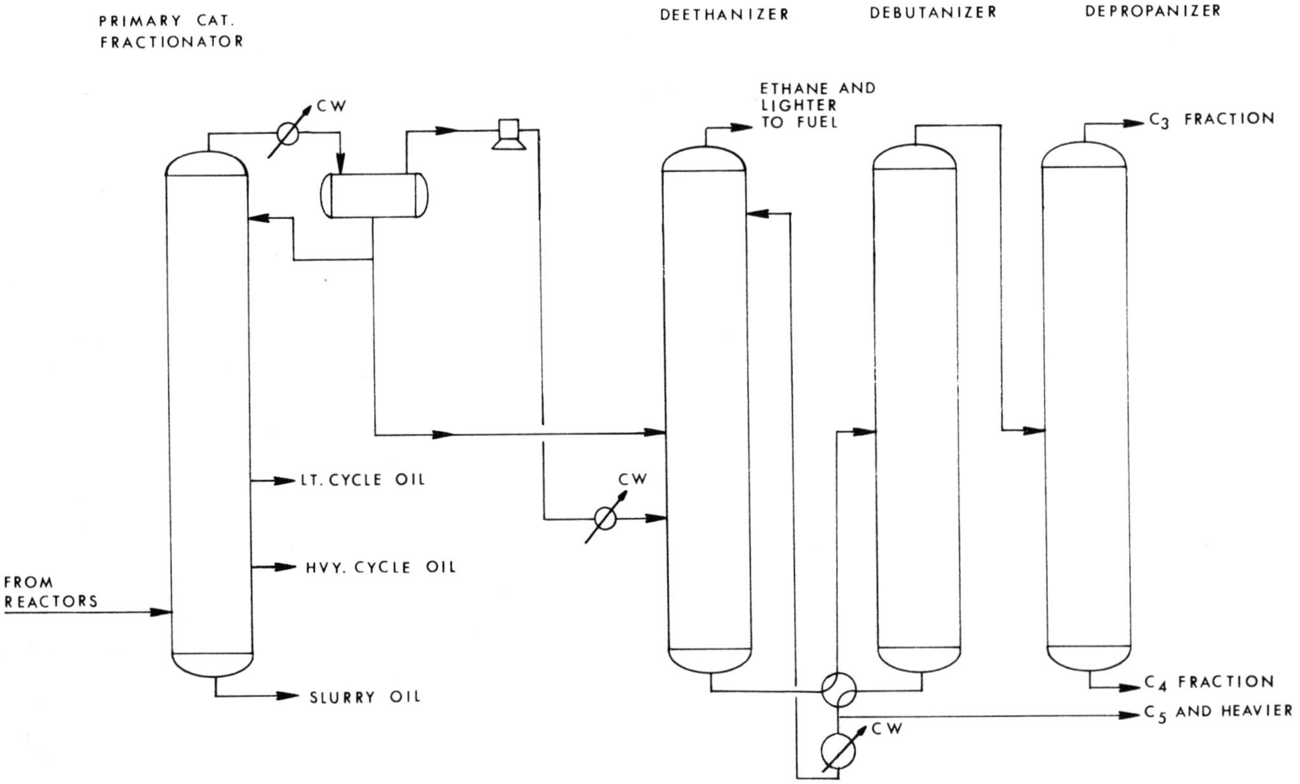

Figure 5.9. FCC unit gas plant for recovering propane and heavier.

degrees F at which temperature it, together with the raw gas condensate, enters the ROD as a two-phase mixture. This stream is quite a bit heavier than the gas fed to the absorber and usually contains 70 to 80 mole percent heavy key and heavier and is 80 to 90 mole percent liquid. As a rich oil deethanizer, this tower will produce ethane and lighter as an overhead gas while recovering propane and heavier as a bottoms liquid. In the FCCU gas plant, the function and operation is generally the same as in the natural gasoline plant. The charge to the FCCU plant is leaner in ethane and lighter, thus allowing operation at a higher temperature and lower pressure level. If these plants had been designed for ethane recovery, the ROD would function as a rich oil demethanizer. ROD bottoms would still flow to the rich oil fracionator whose overhead would now flow to a conventional deethanizer where the ethane, or C_2 fraction, would be yielded as a vapor distillate. The rest of the recovery section would remain unchanged.

Note the presaturator shown in Figure 5.8. The purpose of this is to saturate the lean oil to the high pressure absorber with light components at the pressure level of the ROD. The presaturated liquid is then pumped to the absorber. Since it contains light ends, the absorption of these components in the absorber is considerably less than if pure lean oil had been used. This serves two purposes. It allows a lower absorption temperature and lean oil circulation rate at the same level of refrigeration since the absorption of light components is minimized. But more important from the viewpoint of both capital investment and operating costs, it minimizes the size of the residue gas compressor since the gas which is absorbed in the presaturator is pumped up to the absorber pressure rather than having to be compressed up.

Principles of Design and Operation

The design of most natural gas processing plants is based on the recovery of a certain component, usually either propane or ethane. Since both the absorber exit gas and the ROD overhead gas contain the base component as the heavy key, the design material balance of those two systems must be interdependent. For example, in setting up the preliminary material balance for a gas plant based on 70 percent propane recovery, the absorber must overabsorb so that there is enough extra propane in the ROD feed to permit a reasonably low lean oil rate to the latter tower. The optimum absorption will result in the lowest total lean oil rate to the absorber and the ROD combined. As a rule-of-thumb, the absorption section of the ROD will have to absorb a minimum of 95 percent of the heavy key vapor feed, including the internal vapor.

Design Basis

The overall material balance will usually be set prior to commencing the design of the ROD and will be based on reasonable key component separations. Normally, a minimum of 97 percent of the heavy key is recovered in the bottoms. The maximum amount of light key in the bottoms is often set by downstream product specifications. The operating pressure should be set as high as possible, consistent with using the available heating media. The maximum operating base temperature of the bottoms liquid should not approach the pseudocritical temperature of the bottoms liquid by any closer than 25 degrees F.

Economics

The temperature level chosen for the lean oil will be determined by several factors. Obviously, the lower the temperature, the lower the oil rate and the smaller the reboilers and lean oil chillers. But, on the debit side, lower temperatures increase refrigeration costs, require somewhat more expensive materials of construction and more insulation and suffer more severe startup and operating problems. The lean oil temperature will dictate the overall economics of a given plant design once the key component recovery has been set. This is a detailed subject outside the scope of this work but is mentioned here so that it is well understood that lean oil temperature cannot be determined arbitrarily.

The feed flash temperature also plays an important part in the economics of design. Light ends remaining in the liquid must be stripped out. They carry with them heavy ends which must be reabsorbed. Conversely, the less vapor in the fresh feed, the lower the lean oil rate. Most RODs run at feed temperatures of 80 to 130 degrees F. If the feed is cold, e.g., from a cold absorber, refrigeration should be recovered by cooling hot lean oil. If the feed is hot, e.g., from compressor discharge, it should be cooled to the lowest practical temperature, using either cooling water or air.

Absorption Section Calculations

The calculation procedures discussed in this section are based on estimating the total vapor feed to the absorption section. This is followed by conventional calculations to determine lean oil and tray requirements for the desired separation. The final step calculates the temperature profile across the absorption section and estimates intercooler requirements.

Known Data. The following information must be established firmly prior to commencing calculations.

1. Feed total flow rate and composition.
2. Feed temperature and pressure.
3. Lean oil composition and available temperature levels. Maximum lean oil availability if there is a limitation.
4. Recovery of heavy key in the bottoms liquid and rejection of light key to the overhead gas.

Feed Flash. Calculate the feed flash. Using this and the key split required by the design separation, set up a material balance table and flowsheet showing the internal streams at the feed tray.

Estimate of Key Components Leaving the Stripping Section. The flow rates of the light and heavy key components in the vapor stream leaving the stripping section is estimated by a continuously iterative technique which is based on the assumption that the ratio of heavy key to light key in the stripping section exit vapor, V'_S, is the same as in the feed flash vapor and that the ratio of light key to heavy key in the absorption section exit liquid is the same as in the feed flash liquid. This is the definition of a feed tray from an equilibrium viewpoint. Referring to Figure 5.10, tabulate the following data from the feed flash calculation and the overall tower material balance.

1. Heavy key, moles/hour in feed flash vapor, F'_V, as HKF'_V.
2. Heavy key in overhead gas, G, as HK2.
3. Light key in feed flash liquid, F'_L, and LKF'_L.
4. Light key in bottoms liquid, L, as LK5.
5. Calculate the ratio of heavy key to light key in F'_V and designate as K'_1.
6. Calculate the ratio of light key to heavy key in F'_1 and designate as K'_2. Set up a flowsheet and material balance for the iterative procedure.

For iteration 1, the following arithmetical operations are performed.

1. In stream 1, F_{VA}, the heavy key is the same value as in F'_V. $(HK1)_1 = (HKF'_V)$ where the numeral 1 inside the parentheses refers to the stream number, and the subscript numeral 1 denotes the iteration number.
2. The heavy key in the absorption section downflow liquid, L'_A, is $(HK3)_1 = (HK1)_1 - (HK2)$.
3. The light key in L'_A is $(LK3)_1 = K'_2(HK3)_1$.
4. The light key in the stripping section feed liquid, F_{LS}, is the sum of the absorption section exit liquid and the feed flash liquid. $(LK4)_1 = (LK3)_1 + (LKF'_L)$.
5. The light key in the stripping section overhead vapor, V'_S, is $(LK6)_1 = (LK4)_1 - (LKL)$.
6. The heavy key in V'_S is calculated as $(HK6)_1 = K'_1\ (LK6)_1$.

The value of $(HK6)_1$ is the first estimate of the heavy key from the stripping section. It is added to the heavy key in the feed flash vapor for the second iteration.

Iteration 2 is made as follows.

1. $(HK1)_2 = (HK6)_1 + (HKF'_V)$.
2. $(HK3)_2 = (HK1)_2 - (HK2)$.
3. $(LK3)_2 = K'_2(HK3)_2$.
4. $(LK4)_2 = (LK3)_2 + (LKF'_L)$.
5. $(LK6)_2 = (LK4)_2 - (LKL)$.
6. $(HK6)_2 = K'_1\ (LK6)_2$.

This procedure is repeated until the following is satisfied.

$$(HK6)_{n+1} = (HK6)_n.$$

Nine to 10 iterations will usually suffice to define the total key feed to the absorption section.

Estimate of Average (L/V). The temperature of the gas leaving the top tray will be determined principally by the temperature of the lean oil. The temperature approach at the top of the tower can be anywhere from 5 to 50 degrees F and will depend somewhat on the presence of components lighter than the keys in the exit gas. It is generally true that the average temperature in the absorption section will be no more than 5 degrees F higher than this exit gas temperature. In the case of reboiled absorbers which have liquid intercoolers, the above statement can almost be considered a rule.

The question of trays versus lean oil rate is a classical illustration of economic analysis. However, in the case of reboiled absorbers, the solution is seldom clear-cut since a savings in trays and tower height requires an increased lean oil rate. This in turn increases the refrigeration requirements and necessitates greater capital investment and operating costs. For the most part, the number of trays is preset, allowing lean oil rates to vary as a function of key component recovery. A suggested value for the number of trays is 10 theoretical trays, which will usually result in approximately 20 real trays. For low recoveries, this may result in lean oil rates which are too low for satisfactory tray operation. In this case, the number of trays should be reduced to provide an adequate oil rate.

The L/V ratio for the required recoveries of the key components are calculated for a range of temperatures encompassing the expected absorption temperature. This procedure is as follows.

1. Calculate the required absorption for the total light and heavy key components entering the absorption section as estimated by the iterative procedure.

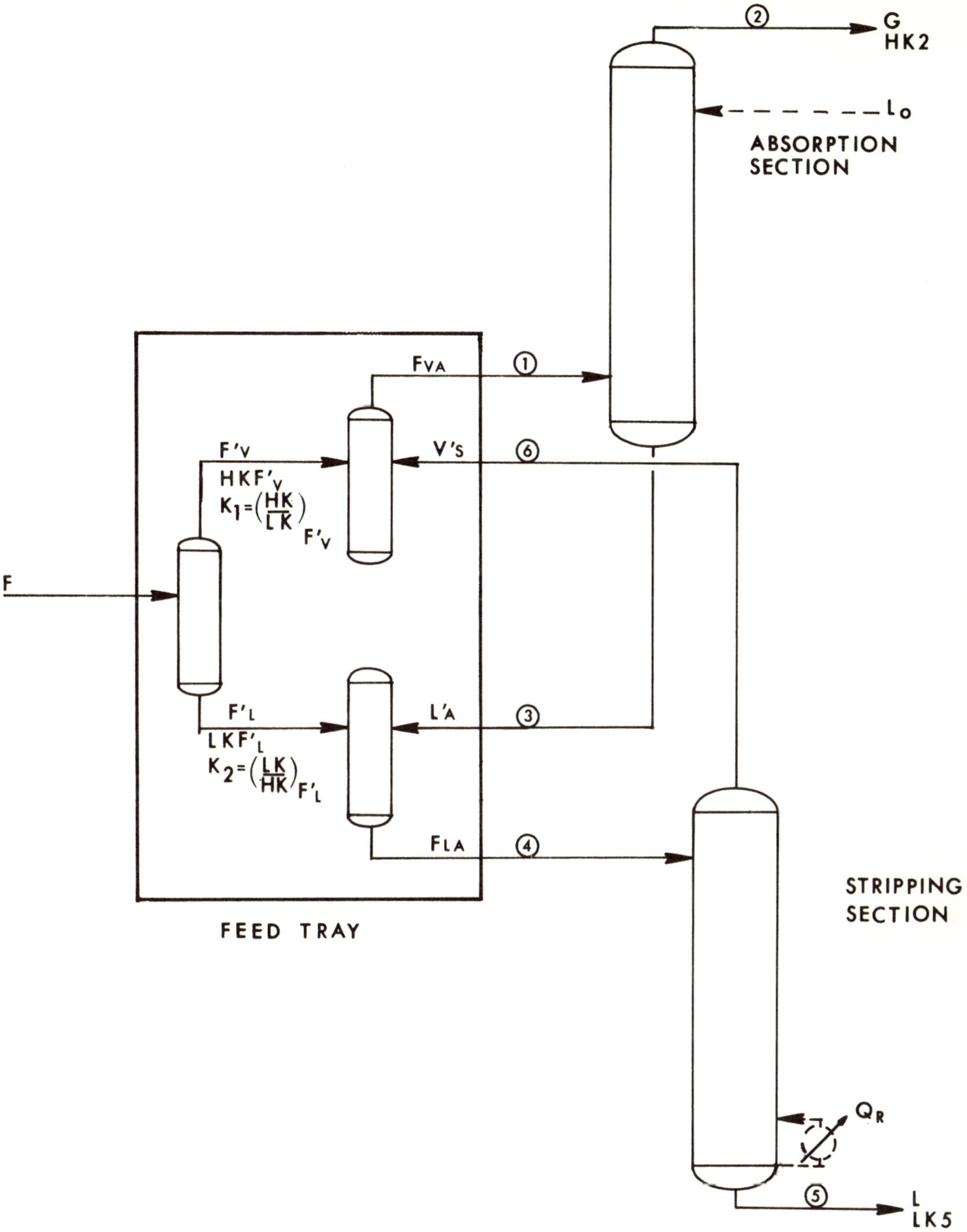

Figure 5.10. Vapor and liquid traffic in a reboiled absorber. Note: lean oil and reboiler not included in overall or feed tray balances.

2. Using an absorption factor chart (20), find the proper values of the absorption factor, A, for 10 trays.
3. Using K values covering the desired temperature range, calculate the required L/V as

$$L/V = AK.$$

The component which has the highest value of L/V is the one whose recovery sets the design.

An interesting point which has come to light in the analysis of several demethanizers and deethanizers is that, in demethanizers, the controlling separation remains between C_2 and C_3 up to a recovery level of about 75 percent, based on fresh feed. Above this level of C_2 recovery, the calculation of a $C_1 - C_2$ split more accurately describes the operation.

Light Ends in Absorption Section Feed Gas. Estimate the rates of components in the feed gas to the absorption section which are lighter than the controlling key component at the same temperature used in the previous step. This procedure is as follows.

1. At a constant absorption factor which is determined by the controlling key, calculate L/V = AK at the indicated temperature levels.
2. Tabulate K values for all light ends at these temperatures.
3. Calculate the absorption factors for all the light ends.
4. From the absorption factor chart, estimate recoveries for the light ends using 10 trays. From these recoveries, the light ends in the feed gas can be estimated.

Ordinarily, the variation in feed gas quantity with temperature will be relatively small. For variations of this order of magnitude, the use of flow rates at a temperature thought to be near the average absorption temperature is justifiable since the equilibrium data are seldom any more accurate than this.

Heavy Ends in Absorption Section Feed Gas. The presence in the absorption section feed gas of components heavier than the controlling key component is estimated by the following assumption.

> Components heavier than the controlling (heavy) key are present in the stripping section exit vapor and, hence, in the absorption section feed vapor, in the same ratio to this heavy key as they exist in the feed flash vapor.

This is expressed algebraically as

$$(V_i)_{FVA} = (V_i)\ F'_V\ [(HK)_{FVA}/(HK)_{F'_V}]$$

Absorption Section Feed Gas. Using the information developed in the previous two steps, tabulate the composition of the feed gas to the absorption section.

Top Tray Calculations. A complete material, heat and equilibrium balance is made around the top tray in the tower by the following step-by-step procedure.

Component Absorption

From the absorption factor chart, estimate the absorption of the heavy key and lighter components on the top tray. Note that the heavy key absorption factor is constant throughout the design. The degree of light ends absorption on the top tray can be estimated closely since their total absorption has already been established.

Estimate of Operating Temperature and Lean Oil Rate

At the same assumed temperature, the lean oil rate required to form a bubble point liquid with the estimated top tray absorbate is calculated. This rate is then used to calculate the *average* L/V and the key component absorption factor.

$$(L/V)_{AVG.} = 0.5\ [(Lo)/G + (Lo + (F_{VA} + G))/(F_{VA})]$$

$$A_{HK} = [(L/V)_{AVG}]\ (K_{HK})$$

where the heavy key K value is evaluated at the assumed temperature. This is a trial-and-error procedure.

Heat Balance

The heat released by the absorbate is calculated as an isothermal condensation at the calculated temperature of the top tray. This heat must be absorbed by the lean oil. If the lean oil rate which was calculated as shown above is too small to absorb the heat, assume a higher top tray temperature and repeat the calculations.

$$t_{Lo} = (Q/Lo)/C_{Lo}, \text{ degrees F.}$$
$$t_{Lo} = (t_{TOP\ TRAY} - t_{Lo\ INLET}), \text{ degrees F.}$$

Lean Oil Losses to the Overhead Gas Stream

The amount of lean oil going overhead with the gas stream is estimated by making a dew point calculation on the exit gas at the top tray temperature.

Total Lean Oil Requirements

The total amount of lean oil which must be fed to the top tray of the tower is the sum of the amounts calculated above.

Tray-by-Tray Material Balance. Using an absorption factor chart and the assumed average absorption temperature, estimate the absorption of each component on each tray and tabulate a tray-by-tray material balance. Note that the components lighter than the heavy key are absorbed from the top while the heavy key and heavier are absorbed from the bottom.

Calculated Temperature Profile. Develop a temperature profile for the absorption by calculating bubble points on the liquid leaving each tray. Plot this temperature profile and calculate the average temperature as the arithmetic average of all points. This is the average absorption temperature.

Estimate the heat released on each tray by the absorbate. Calculate the lean oil temperature rise across each tray. Plot this temperature profile and calculate the average absorption temperature. In many cases, these temperatures will be somewhat greater than those calculated from bubble points.

If the average absorption temperature as calculated from bubble points exceeds the assumed average absorption temperature, use intercooling of lean oil back to its base temperature. The appropriate tray for locating the intercooler or intercoolers if more than one is required can be determined by inspection of the temperature profiles.

Plot a new temperature profile using the heat release data but showing the effect of the intercooler(s). This new average absorption temperature is the final value and must not exceed the orginally assumed value.

Calculate the duties of the intercooler(s).

Stripping Section Calculation

Detailed instructions for calculating the stripping section of the tower and the overall heat balance are not given here since they are quite straightforward.

References

1. W.L. Nelson, *Petroleum Refinery Engineering*, 4th ed. (New York: McGraw-Hill Book Company, Inc., 1958).
2. R.J. Hengstebeck, *Petroleum Processing* (New York: McGraw-Hill Book Company, Inc. 1959).
3. J.B. Maxwell, *Data Book on Hydrocarbons* (Princeton, N.J.: D. van Nostrand Co., 1965).
4. W.C. Edmister, *Applied Hydrocarbon Thermodynamics* (Houston: Gulf Publishing Company, 1961).
5. *Technical Data Book–Petroleum Refining* (Washington, D.C.: American Petroleum Institute, 1966).
6. J.M. Ascrizzi, "Ethane Specifications for Ethylene Production," *Hydrocarbon Processing* 46, no. 4, (April, 1967), pp. 139-40.
7. W.D. Harbert, "Which Tower Goes Where?" *Petroleum Refiner* 36, no. 3 (March, 1957), pp. 169-74.
8. H.C. Bozeman, "New Techniques Spur Changes in Light Ends Processing," *Oil and Gas Journal* (August 31, 1964), pp. 48-57.
9. W.L. McCabe and E.W. Thiele, "Graphical Design of Fractionating Columns," *Industrial and Engineering Chemistry* 17, no. 6 (June, 1925), pp. 605-11.
10. A.J.V. Underwood, *J. Inst. Pet.* 32 (1926), pp. 614-26.
11. M.R. Fenske, *Industrial and Engineering Chemistry* 24 (1932), p. 482.
12. F.W. Winn, "New Relative Volatility Method for Distillation Calculations," *Petroleum Refiner* 37, no. 5 (May, 1958).
13. E.R. Gilliland, *Industrial and Engineering Chemistry* 32 (1940), p. 1220.
14. W.A. Mason, *Petroleum Refiner* 38, no. 5 (May, 1959), p. 237.
15. W.C. Edmister, "Absorption and Stripping Factor Functions for Distillation Calculations by Manual and Digital-computer Methods," *AIChE Journal* 3, no. 2 (June, 1957), pp. 164-71.
16. R.J. Hengstebeck, "A Simplified Method for Solving Multicomponent Distillation Problems," *AIChE Transactions* (1946), pp. 309-29.
17. R.J. Hengstebeck, "A Rigorous Graphical Method for Calculating Multicomponent Distillations," *Chemical Engineering Progress* 53, no. 5 (May, 1957) pp. 243-48.
18. C.S. Robinson and E.R. Gilliland, *Elements of Fractional Distillation* 4th ed. (New York: McGraw-Hill Book Company, Inc., 1950).
19. B.D. Smith, *Design of Equilibrium Stage Processes* (New York: McGraw-Hill Book Company, Inc., 1963).
20. *Engineering Data Book* (Tulsa, Oklahoma: Natural Gas Processors Suppliers Association, 1966).

pyrolysis gas quench tower

Although a pyrolysis gas quench tower physically appears to be very similar to other types of refinery fractionators considered previously, there is very little process similarity. Strictly speaking, a quench tower is a direct-contact gas cooler and scrubber, and any separation that occurs is limited to a single stage flash. However, this type of tower is included for discussion because the general calculation procedure is similar to those used elsewhere and because the analysis of the pyrolysis oil separation is handled most easily by using petroleum oil techniques.

This chapter describes a pyrolysis gas quench tower for a light naphtha cracking plant in ethylene production. The tower, as illustrated in Figure 6.1, is designed to employ a two-stage oil quench operating with indigenously produced quench liquids. The attendant water section of the tower is a single stage operation. Also included are general procedures for calculation of a tower operating with an outside quench oil and for calculation of an all-water tower.

These techniques may be applied to any type of quench tower design, whether it be propane cracking for ethylene or vacuum region deep cracking such as for acetylene or butadiene manufacture, as long as the proper analysis and characteristics of the pyrolysis oils are correctly identified.

While a quench tower has trays or packed sections, external coolers, reflux lines and all the apparent paraphernalia of any conventional refinery distillation column, it does not do any fractionation. Any interphase mass transfer occurs by virtue of being cooled below its dew point by countercurrent contract with cool oil or water. All process heat is removed from the system by cooling the circulating quench fluids in external exchange equipment. No significant amount of process heat is removed by internal reflux within the tower. The reasons for this are fairly obvious. Hot, spent quench liquid is pumped from the bottom of a section, through a cooler and back to the top of the same section. From an equilibrium viewpoint this liquid is always much heavier than it should be with respect to the vapor composition of the zone under study and can be considered an equilibrium liquid only upon leaving the bottom tray. This greatly simplifies the calculations since all separations can be considered as simple flashes, thus avoiding tedious fractionation calculations.

Properties of Tower Feedstream

The effluent gas from a cracking furnace consists of a large volume of light hydrocarbon gases, the dilution steam and a small quantity of heavy pyrolysis products which are liquids at ambient temperature and atmospheric pressure. The relative amounts of pyrolysis oil, to be referred to as PO, are a function of the feedstock to the furnace and cracking severity, i.e., depth of cracking. The conditions of the feed to any quench tower in terms of composition, temperature and pressure are set by two factors. The first is the furnace outlet temperature, determined by reaction conditions which are normally in the range of 1,500 degrees F. The second is the amount of cooling that takes place between the furnace and the tower. Current designs call for generating high pressure superheated steam immediately downstream of the furnaces followed by line quenching. This latter cooling is accomplished by directly injecting oil or water into the cracked gas transfer line. The principal aim is to reduce the cracked gas temperature as rapidly as possible to avoid further uncontrollable reactions. The heat recovery available from this cooling is of secondary importance to the process, although it may be a great overall significance to the economics of the total plant. From this discussion, it becomes obvious that the temperature of the cracked gas entering the quench tower will be determined independently of any tower design considerations.

The breakdown of the hydrocarbon portion of the furnace outlet into gas and heavy oil components is due to the sampling methods of a furnace effluent. The hot gas is cooled in a sampling device, usually nothing more than a small water cooled heat exchanger. The resulting gas is separated from the liquids produced, caught in a sample bomb and analyzed by an appropriate method. The condensate, mostly water, is extracted, and the hydrocarbon portion of the liquids formed is recovered from the extract and analyzed. While it would be possible to synthesize a continuous TBP curve to describe both the gas and heavy oil fractions, it serves no useful purpose since fractionation does not occur here. The general characteristics of the portions of the feed are given below.

Hydrocarbon Gas

The composition of this stream is given on a weight and a molar basis. It will normally cover all light compounds present from hydrogen through C_4 or C_5, which represents the cut point between the gas and the residual liquid fractions. For this portion of the total tower feed, calculate the average molecular weight, critical temperature and pressure and the molal average (60 degrees F, 14.7 psia) heat capacity.

Pyrolysis Oil

The pyrolysis oil, PO, is characterized as a conventional full-range hydrocarbon liquid, the nature of which is somewhat different than those normally encountered in a refinery. While it is similar to streams in a catalytic cracking unit in that it is rich in olefins, diolefins and aromatics, it is much more highly cracked. Thus, the heavier portion which boils above approximately 425 degrees F is rich in polynuclear aromatics such as naphthalene and similar materials. In some cases, a solid phase may develop upon cooling the heavier portions. The process designer must consider such

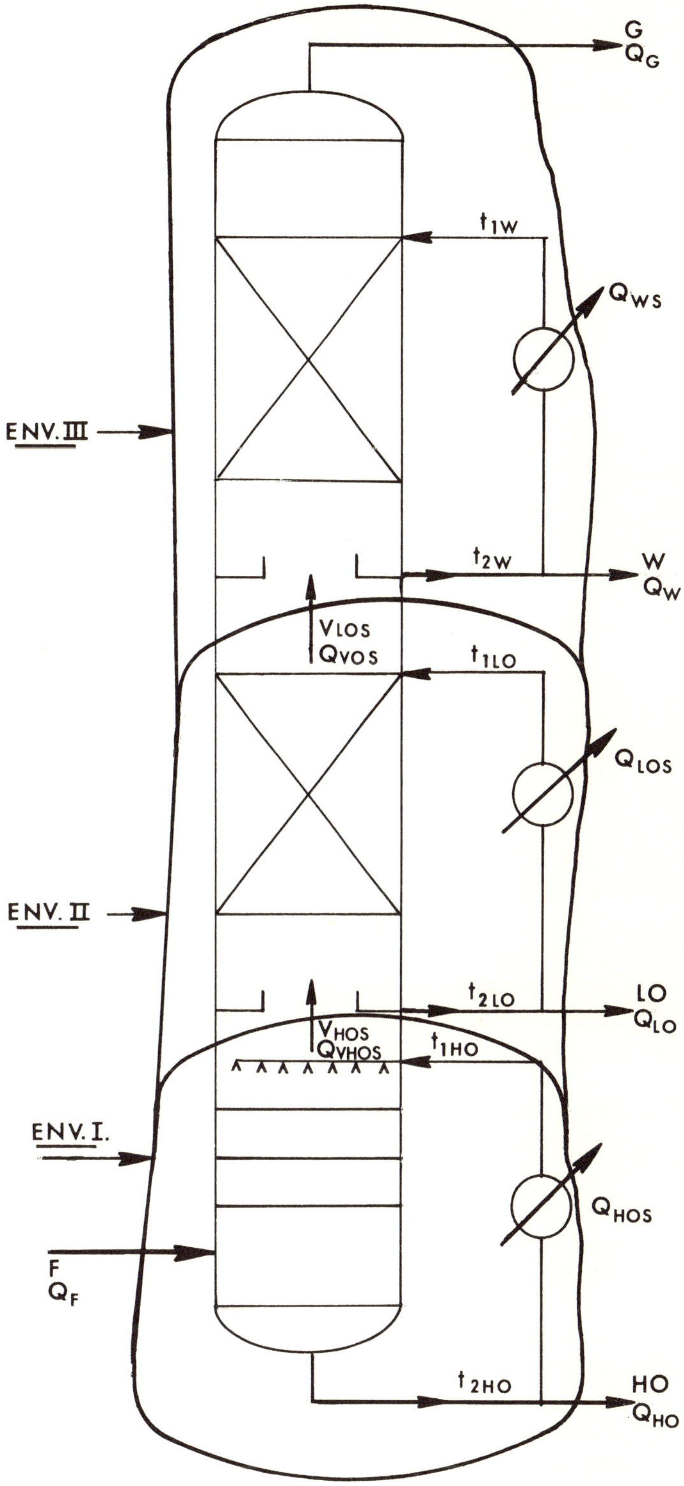

Figure 6.1. Pyrolysis gas combination—oil and water quench tower.

possibilities and obtain experimental data describing the properties of the oil under study. For the purposes of separation calculations, the TBP distillation and gravity of PO are required as basic data. From this, the following data should be developed:

a. Atmospheric EFV curve by Packie's procedure.
b. Phase diagram.
c. Vacuum region phase diagram by Edmister's procedure.
d. Weight-volume-mole relationship.

Steam

This is the dilution steam that is charged to the furnace along with the cracking charge stock. The steam dilution rate is set by reaction section design requirements and is usually described as pounds steam per pound of charge.

General Basis of Design

In a combination oil-water quench tower, the hot cracked gas is first contacted with cool oil which reduces the gas temperature to a point slightly above the water dew point of the gas and at the same time condensing out the heavy portion of the PO. The gas then passes to the water section where it is cooled by direct water washing. The dilution steam and some additional lighter PO are condensed. The general points to be considered in setting up the design are as follows.

Oil Section

Pyrolysis oil is a valuable liquid rich in benzene, toluene, xylenes, naphthalene and other materials that are in high demand. Unfortunately, the heavy high boiling back end has little material value for other than fuel oil. It can be converted to a satisfactory catalytic cracking feedstock by severe hydrotreating, and it can be hydrocracked. However, even in a large cracker, the volume of PO produced is so small that it could not support such processing units by itself. Even so, it is advantageous to operate a two-stage oil quench section if the total yield of oils in these two sections would be sufficient to justify splitting them. The lower section would be designed to condense out everything heavier than catalytic heavy-cycle oil which requires a TBP cut point of 850 to 950 degrees F. By using a high scrubbing liquid rate, the entrainment of heavy oil and/or pyrolysis semisolids to the upper oil section is minimized. The remainder of the oil condensed out at the conditions entering the water section will be recovered in the upper section and should be relatively clean.

This type of design is advisable only if oils are to be recovered. In this case, the upper section oils are sent to a hydrotreater, an aromatics recovery unit, a coker or a similar unit. It is not suitable for cat-cracking feedstock because of its high aromaticity. The oil from the lower section is routed directly to fuel oil blending to a residual hydrotreater or to a coker if available.

If the recovered oils are used only for fuel, the more economical design would be only one oil section, and the liquid would consist of the total oil condensed at the temperature and pressure conditions entering the water section.

From an equipment viewpoint, the one oil section design is more economical since only one set of pumps, exchangers and piping are required. However, the two-oil section design provides two levels of hot oil temperatures, giving a higher degree of heat recovery or possibly low pressure steam generation. The design of the oil section should give thorough consideration to the various questions of incremental capital investment and operating costs.

Water Section

The function of the water section is to cool the gas from the oil section exit temperature to the minimum possible temperature consistent with the available utility cooling medium, simultaneously condensing the bulk of the dilution steam. It is quite possible to achieve a 5 degree F approach of cooled gas to inlet quench water temperature. This is a very important point since the lower the quenched temperature, the lower the horsepower required for the cracked gas compressor. It has generally proven economical to design the external quench water coolers for as low as a 10 degree F cold end approach. This, in effect, means that the design temperature of the quenched cracked gas should be no less than 15 degrees F higher than the maximum available cooling water temperature.

The inlet temperature of gas to the water section should be approximately 50 degrees F above the cracked gas water dew point temperature. The cool oil temperature to the top tray of the upper oil section should be set at least 15 degrees F above the dew point. This latter consideration cannot be violated without the inherent possibility of liquid water condensing in the oil section causing severe foaming of the oil and affecting tower stability.

The temperature of the outlet water is open to wide choice by the designer. The water rate can be minimized at the expense of heat transfer trays or vice versa. At this point, refer to Figure 6.2 which is a typical cracked gas cooling curve. A line drawn from the inlet water point, coordinates (zero BTU per hour, 95 degrees F), through the water dew point on the cracked gas curve, coordinates—(129.3 million BTUs per hour, 195 degrees F), and extended to the water section limit line, approximate coordinates (136 million BTUs per hour, 201 degrees F), describes the minimum water rate and is analagous to a minimum reflux rate in distillation work. This means that, for this case, the quench water must leave the tower at a temperature of less than 201 degrees F in order to absorb the

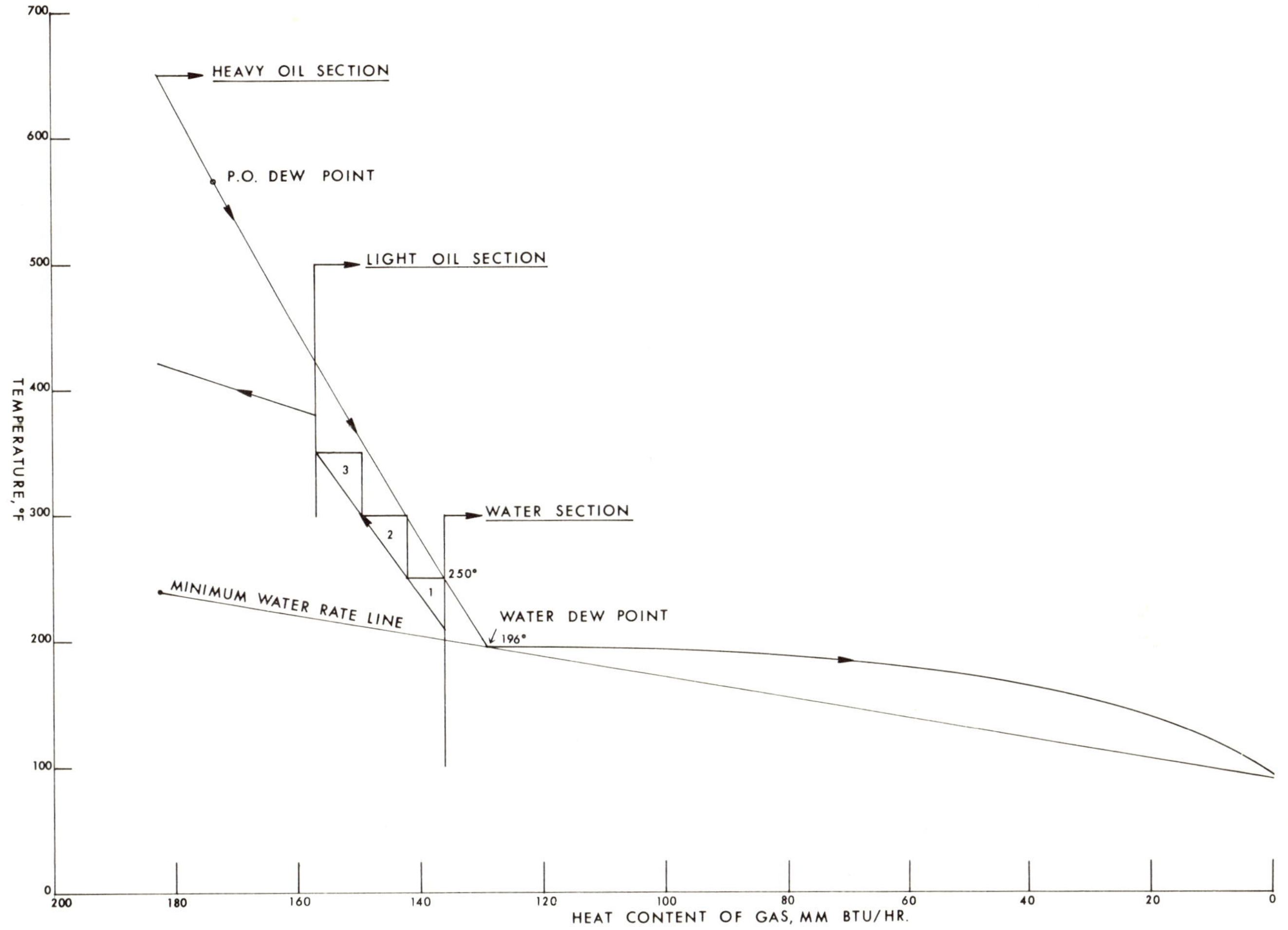

Figure 6.2. Typical cracked gas cooling curve and quench tower temperature profile.

required amount of heat, using a finite number of heat transfer trays.

The quench water usually flows to a fairly large settling vessel operated at atmospheric pressure. In this vessel, a separation occurs between the quench water and the oil condensing in the water section. Limiting the hot quench water to a maximum of 180 degrees F provides for adequate suction pressure to the quench water pumps. In Figure 6.2 the outlet water temperature was set at 150 degrees F because of the pinch which could have developed had 180 degrees F been selected. On the surface of it, this seems like a good subject for an economic study, but without good heat transfer tray data, accurate results cannot be attained.

Tower Pressure Drop

The basic operating principle of the quench tower is to cool the gas to the minimum temperature feasible with the available cooling medium. This will guarantee the maximum possible condensation. However, the gas must be thoroughly scrubbed in order to wash back the nearly infinitesimal carbon and pyrolysis solids particles. This can only be accomplished by bringing the gas and liquid into vigorous contact. As is known from tray design principles, contacting efficiency can be obtained only at the expense of pressure drop. In this respect, a quench tower can be thought of as an absorber as well as a gas cooler. For this reason, it is recommended that the tower be designed for a maximum pressure drop of 5.0 psi for pressure services. For vacuum quench towers where this much pressure drop is not available, the problem can be overcome by using a grid-type packing which provides good contact with reasonably low pressure loss.

Heat and Material Balance Calculations

The following outlines a procedure for calculating the heat and material balance around the various sections of a combination oil-and-water quench tower.

Preliminary Calculations

The following preliminary calculations are required to establish the basic design approach.

1. At the tower inlet pressure, calculate the PO dew point temperature. Calculate the water dew point temperature assuming no previous condensation of PO.
2. Given the overhead temperature and total pressure, calculate the quantity and composition of PO which remains vaporized. This calculation is made as follows.
   a. The total gas leaves the tower saturated with water vapor. Calculate the mole fraction of (hydrocarbon gas + PO) leaving the tower in the overhead gas.
   b. Assume a volume percent vaporization and convert to moles per hour PO vapor going overhead by using the volume-mole-weight curves.
   c. Calculate the PO partial pressure, P'_{PO}, in the overhead vapor.
   d. At P'_{PO} and the vapor temperature, find the volume percent vapor at point (P'_{PO},T_V) on the vacuum region phase diagram. This point is translated to the atmospheric EFV curve by assuming that the difference between T_V and the nearest line of constant vaporization is constant for the EFV curve at atmospheric pressure and below. The vaporization on the EFV curve should check the assumed value. If it does not, repeat the procedure.
3. Having set the temperature and pressure of the inlet gas to the water section, make a flash calculation on the PO at that point by the same procedure as in Step 2 above.
4. Set the temperature and pressure at the light oil section inlet. This is determined by deciding what portion of the oil is to be condensed in the heavy oil section. The criteria for this decision were discussed in the section, "General Basis of Design-Oil Section." Note that this is also the bottoms liquid temperature. Calculate the flash on PO at these conditions by the same procedure as in Step 2 above.
5. Calculate the moles per hour, pounds per hour and gallons per hour of the oil yielded from all three sections and from the overhead stream.
6. Set up a process flowsheet and material balance for all inlet and outlet streams. Record key temperatures for the gas and the bottoms liquid.

Gas Cooling Curve

In order to determine the heat effect across each section of the tower, it is useful to calculate and to plot a cooling curve for the cracked gas, taking into account the condensation of water and PO. This curve is required for the thermal design of the tower.

PO flash calculations have already been made in the preliminary calculations. Thus, it is simple to calculate the heat curve from the tower inlet temperature down to the water section inlet temperature. In the water section, it is necessary to set several temperatures and pressures, calculate the amount of water condensed and then calculate the heat effect, evaluating the heat content of the condensate at the outlet quench water temperature. Note that the water section calculations are based on assuming that the pressure drop and molal PO condensation will vary linearly with temperature.

External Cooling Streams

Water Section

The minimum inlet water temperature is set by assuming an approach of 10 degrees F to the available cooling water temperature. The criteria for setting the exit water temperature has already been covered.

Light Oil Section

The minimum oil inlet temperature should be at least 15 degrees F above the water dew point of the gas. The outlet oil temperature should be set to provide a lower heat level than is available from the base section. A rough rule-of-thumb is that the hot light oil should be no closer than 50 degrees F to the hot heavy oil temperature. This will result in an effective two-temperature level heat recovery system. A secondary procedure for setting the hot oil temperature is to select a hot temperature point which will yield approximately three McCabe-Thiele-type heat transfer stages as shown for the light oil section on Figure 6.2.

Heavy Oil Section

The temperature of the outlet heavy oil is the same as the bottoms temperature. The cool oil temperature is determined by the oil rate which is a function of tower diameter. This must be set at the last of the design. It is desirable to have a cool heavy oil at a higher level than the hot light oil temperature.

Heavy Oil Section Heat Balance

The heavy oil section is calculated by making a heat balance around Envelope I as shown on Figure 6.1.

$$Q_{IN} = Q_{iHOS} = Q_F$$

$$Q_{OUT} = Q_{OHOS} = (Q_{HO} + Q_{VHOS})$$

The external heat removal requirement is calculated by

$$Q_{HOS} = (Q_i - Q_o)_{HOS}$$

Light Oil Section Heat Balance

The light oil section is calculated by making a heat balance around Envelope II of Figure 6.1.

$$Q_{IN} = Q_iLOS = Q_F$$

$$Q_{OUT} = Q_{OLOS} = (Q_{HO} + Q_{LO} + Q_{HOS})$$

The external heat removal requirement is calculated by

$$Q_{LOS} = (Q_i - Q_O)_{LOS}$$

The circulating oil rate in the light oil section is calculated by

$$W_{LOS} = [Q_{LOS}/(h_2 - h_1)_{LO}]$$

Water Section Heat Balance

The water section is calculated by making a heat balance around Evelope III of Figure 6.1.

$$Q_{IN} = Q_{iWS} = Q_F$$

$$Q_{OUT} = Q_{OWS} = (Q_W + Q_{LO} + Q_{HO} + Q_{HOS} + Q_{LOS}).$$

The external heat removal is calculated by

$$Q_{WS} = (Q_i - Q_O)_{WS}$$

The circulating water rate in the water section is calculated by

$$W_{WS} = [Q_{WS} / (h_2 - h_1)_{H_2O}]$$

Operation with Quench Oil Makeup

In the event that the rate of PO condensation in the system is insufficient to keep tars and heavy ends from accumulating to an inoperable degree, an external oil may be fed to the tower at a suitable rate. In designing for this case, it is necessary to flash the external oil at the exit conditions from the various sections, actually reducing to a double trial-and-error flash on both PO and the outside oil. But since very small amounts of material are present with respect to the large amounts of gases and water vapor, the calculation is easily resolved.

The type of oil for this service is very important. Since the liquid is in contact with hot gases at relatively low pressure, the quench oil will decompose if it is a suitable cracking feedstock. Thus, straight-run distillates or gas oils should not be used for this service. Rather, a catalytic cycle oil, a coker gas oil or a stream with a similar processing history should be used.

All-Water Quenching Operation

In cracking plants designed for light hydrocarbon feedstocks, the volume of PO in the furnace effluent is quite small so that an oil quench is not usually employed. An all-water system is common although it does not do as good a job of washing the gas. In cracking units having an all-water quench system, the downtime for cleaning the cracked gas compressor is significantly greater than in plants using an oil wash. Reboiler fouling problems are also more severe.

The calculations for such a system follow the same general procedure as for the combination tower and is outlined below.

1. Calculate the PO and water leaving with the overhead gas.
2. Set the outlet water temperature.
3. Calculate and plot the cooling curve, evaluating the heat content of the liquids formed at the outlet water temperature.
4. Calculate the required outside heat removal requirements.

Preliminary Process Design Calculations

To this point, all the process design has been done with the exception of establishing the circulating oil rate and cool oil temperature in the heavy oil section. For the purpose of getting a vigorous wash of the hot gas, an inlet oil rate of 20 gpm per square foot of tower cross-sectional area is arbitrarily recommended. There is nothing magic about this rate, but it has been used in the design of many successful catalytic cracking fractionators in the desuperheating and condensing sections, and the services are quite similar. The thermal design of the tower must also be checked and the proper number of heat transfer trays provided. The following procedure is used to arrive at the final process design of the tower.

1. Calculate and tabulate the vapor and liquid loadings at the inlet and outlet of the light oil and water sections.

2. Using a suitable tower-sizing procedure, calculate the required tower diameter at 24-inch tray spacing for the above loadings. By varying tray spacing in all sections as may be indicated, select the optimum minimum tower diameter.
3. Using this tower diameter, set the cool oil rate to the tower at 20 gpm per foot of total cross-section area. For this rate, calculate the cool oil inlet temperature which will absorb the required amount of heat. Tabulate vapor and liquid loadings for the heavy oil section. Calculate the tower diameter for a 24-inch tray spacing.

If this calculated tower diameter is less than the key tray tower diameter at 24-inch tray spacing, the design is satisfactory. If it is greater, the heavy oil section will be the key section of the tower, and the oil rate should be designed for this latter diameter.

Thermal Design

The heat transfer tray area required to absorb the calculated amount of heat is determined by the following procedures:

1. Calculate the log mean temperature difference for the section.
2. From the heat balance calculations, tabulate Q for the section.
3. Using a suitable heat transfer coefficient (References 7 and 8 in Chapter 4), calculate the required tray area and number of trays.

Note that in the water section, there is sufficient deviation from linearity to preclude the use of the log mean temperature difference. The curve is broken down into smaller increments where the log mean temperature difference can be applied.

Vapor-Liquid Traffic

In performing the process and mechanical design of the tower, it is useful to know the vapor and liquid loadings at all key trays. Also, the rapid shrinkage of the vapor in the water section is of particular interest in optimization of the tray design. The data for calculating these loadings is available in the cooling curve calculations and from the water circulation rate. The vapor temperature is then related to the actual tray number of the tower.

crude assay

CRUDE: **TIA JUANA LIGHT**

COUNTRY: VENEZUELA

Representative of: Current (March 1969) grade of Tia Juana Light from the La Salina terminal in Lake Maracaibo. Assay run on two-drum sample received from Amuay.

File No:	SL.121C-AN.69
Report Date:	December 18, 1969
Date Received:	May 1969
Date Distilled:	August 1969
Cost Center:	2520-311

Assay Run By: Humble Oil & Refining Company
Refining Department
Refinery Laboratory
Baytown, Texas

Sponsored By: Esso International – Cargo Sales

Esso International – Supply and Transportation

SO(NJ) – Logistics Coordination Department

SO(NJ) – Planning Coordination Department

E. R. & E. – Engineering Information Center

TABLE 1

CRUDE	
TIA JUANA LIGHT	SL.121C-AN.69

WHOLE CRUDE DATA

Gravity	°API	31.6
Specific Gravity	60/60	0.8676
Sulfur	Wt. %	1.08
Pour Point	°F	-5
Water and Sediment	Vol. %	0.0
Salt Content, NaCl	ptb	0.85
Reid Vapor Pressure	psi	4.9
H_2S (Dissolved)	ppm	0
Neut. No. (D664)	mg KOH/gm	0.11
Viscosities:		
Kinematic @ 70°F	cs	16.1
100°F	cs	10.2
Saybolt Universal @ 70°F	sec.	81.4
100°F	sec.	59.5

LIGHT HYDROCARBONS

% ON CRUDE	Wt.	Vol.
Methane		
Ethane	0.02	0.04
Propane	0.22	0.37
Iso Butane	0.18	0.27
Normal Butane	0.60	0.89
Iso Pentane	0.56	0.77
Normal Pentane	0.82	1.13

TABLE 2

CRUDE	
TIA JUANA LIGHT	SL.121C-AN.69

PERCENT DISTILLED VS 15/5 ASSAY STILL TEMPERATURE (FAHRENHEIT)

15/5 Cut Point °F VT	Total Distilled – % Volume On Crude									
	0	10	20	30	40	50	60	70	80	90
								1.6	2.0	2.4
100	2.8	3.2	3.7	4.1	4.6	5.2	5.9	6.6	7.3	8.0
200	8.8	9.7	10.5	11.3	12.2	13.0	13.9	14.8	15.7	16.7
300	17.7	18.6	19.5	20.4	21.3	22.3	23.2	24.2	25.1	26.1
400	27.0	27.9	28.9	29.9	30.8	31.8	32.8	33.7	34.7	35.7
500	36.6	37.5	38.5	39.5	40.4	41.4	42.3	43.3	44.3	45.3
600	46.3	47.3	48.3	49.2	50.2	51.1	52.0	52.9	53.9	54.8
700	55.7	56.7	57.7	58.6	59.4	60.3	61.3	62.2	63.0	63.8
800	64.6	65.4	66.3	67.2	68.0	68.8	69.6	70.4	71.2	71.9
900	72.6	73.3	74.1	74.8	75.5	76.2	76.8	77.5	78.2	78.7
1000	79.3	79.9	80.5	81.1	81.6	82.1				

TABLE 3

CRUDE	
TIA JUANA LIGHT	SL.121C-AN.69

LIGHT ENDS – COMPOSITION AND CALCULATIONS

	ASSAY STILL GAS 1.84 Liq. Vol. % on Crude			C_5/158°F. VT NAPHTHA 3.97 Liq. Vol. % on Crude		TOTAL LIGHT ENDS	G. C. ANALYSIS OF WHOLE CRUDE Liq. Vol. %
	Mol. % From M. S.	Liq. Vol. % On Cut	Liq. Vol. % On Crude	Liq. Vol. % From G. C. On Cut	Liq. Vol. % From G. C. On Crude	Liq. Vol. % On Crude	
C_2 & Lighter	3.1	2.2	0.04			0.04	0.1
C_3	22.4	20.1	0.37	0.1		0.37	0.4
i-C_4	12.9	13.6	0.25	0.3	0.02	0.27	0.4
n-C_4	42.0	42.9	0.79	2.4	0.10	0.89	0.8
						(1.57)	(1.7)
i-C_5	11.1	13.0	0.24	13.4	0.53	0.77	0.9
n-C_5	6.9	8.2	0.15	24.7	0.98	1.13	1.2
C_6 +	1.6			59.1	2.34	2.34	
Total	100.0	100.0	1.84	100.0	3.97	5.81	

TABLE 4

CRUDE	
TIA JUANA LIGHT	SL.121C-AN.69

STRAIGHT RUN DEBUTANIZED GASOLINES

15/5 Cut Point	°F VT	68-158	68-212	68-248	68-302	68-374
15/5 Cut Point	°C VT	20-70	20-100	20-120	20-150	20-190
Yield Cut Range	Vol. %	1.6-5.9	1.6-9.8	1.6-12.9	1.6-17.7	1.6-24.6
Yield on Crude	Vol. %	4.3	8.2	11.3	16.1	23.0
Gravity	°API	83.5	74.3	69.8	64.8	59.5
Specific Gravity	60/60	0.6581	0.6876	0.7029	0.7208	0.7408
Total Sulfur	Wt. %	0.008	0.008	0.008	0.008	0.011
Mercaptan Sulfur	ppm	2	2	2	2	2
Reid Vapor Press.	psi	10.8	7.4	5.8	4.4	3.3
Research Oct. No.						
Clear		68.0	63.3	60.2	55.7	49.8
+ 1.5 ml TEL/USG		83.2	78.8	75.7	71.2	65.1
+ 3.0 ml TEL/USG		87.8	83.6	80.7	76.7	71.5
Motor Oct. No.						
+ 3.0 ml TEL/USG		87.0	82.7	80.2	77.0	73.2

ASTM D-86*

Vol. % D + L @ 70 °C/158 °F	99.0	35.0	16.0	9.0	4.0
@ 100 °C/212 °F		96.0	69.0	45.0	24.0
@ 125 °C/257 °F			100.0	81.5	46.5
@ 150 °C/302 °F				100.0	72.5
@ 180 °C/356 °F					99.0

Detailed ASTM D-86 data in appendix table 6

TABLE 5

CRUDE	
TIA JUANA LIGHT	SL.121C-AN.69

NAPHTHAS

15/5 Cut Point	°F VT	68/158	158/212	212/248	212/302	248/302	302/3
15/5 Cut Point	°C VT	20/70	70/100	100/120	100/150	120/150	150/1
Yield Cut Range	Vol. %	1.6-5.9	5.9-9.8	9.8-12.9	9.8-17.7	12.9-17.7	17.7-
Yield on Crude	Vol. %	4.3	3.9	3.1	7.9	4.8	6.9
Mid-Point	Vol. %	3.8	7.9	11.4	13.8	15.3	21.
Gravity	°API	83.5	65.0	58.7	55.8	54.1	49.
Specific Gravity	60/60	0.6581	0.7201	0.7440	0.7555	0.7624	0.7
Total Sulfur	Wt. %	0.008	0.008	0.008	0.008	0.008	0.0
Mercaptan Sulfur	ppm	2	2	2	2	2	2
Reid Vapor Pressure	psi	10.8	3.3	1.6	1.0		
Research Octane Number							
Clear		68.0	58.1	51.7	47.8	45.4	39.
+ 3.0 ml TEL/USG		87.8					
Color	Saybolt		+30		+30		+30
Aniline Point	°F		132	128	127	127	132

		GC	GC	HC*	GC	HC*	HC*	HC
Paraffins	Vol. %	90.4	62.6	52.3	52.3	51.9	51.6	47.5
Naphthenes	Vol. %	8.1	32.8	37.2	34.3	34.9	33.3	32.
Aromatics	Vol. %	1.5	4.6	10.5	13.4	13.2	15.1	20.

* Calculated from type analyses data in Table 9.

TABLE 6

CRUDE	
TIA JUANA LIGHT	SL.121C-AN.69

HYDROCARBON COMPONENT ANALYSIS

15/5 Cut Point	°F VT	Gas to 68	68 – 158
15/5 Cut Point	°C VT	Gas to 20	20 – 70
Yield Cut Range	Vol. %	0.0-1.84	1.84-5.81
Yield On Crude	Vol. %	1.84	3.97

	Mass Spec. Liq. Vol. %		Gas Chrom. Liq. Vol. %	
	On Naph.	On Crude	On Naph.	On Crude
Ethane	2.2	0.04		
Propane	20.1	0.37	0.1	0.01
Iso Butane	13.6	0.25	0.3	0.01
n-Butane	42.9	0.79	2.4	0.10
Iso Pentane	13.0	0.24	13.4	0.53
n-Pentane	8.2	0.15	24.7	0.98
2,2-Dimethylbutane			0.9	0.04
2,3-Dimethylbutane			2.1	0.08
2-Methylpentane			16.0	0.64
3-Methylpentane			9.8	0.39
Cyclopentane			2.4	0.10
n-Hexane			20.2	0.80
2,2-Dimethylpentane			0.1	0.01
2,4-Dimethylpentane			0.2	0.01
Methylcyclopentane			4.7	0.19
2-Methylhexane			0.1	0.01
2,3-Dimethylpentane				
3-Methylhexane			0.1	0.01
Cyclohexane			1.0	0.04
Benzene			1.5	0.06

Summary:		
Paraffins	Vol. %	90.4
Naphthenes	Vol. %	8.1
Aromatics	Vol. %	1.5

Actual yields and inspections on fractions shown in Tables 6 and 7 are those as cut from the assay still. Value shown in Tables 4 and 5 differ slightly as they have been adjusted to reflect calculated complete split between C_4 and C_5 hydrocarbons.

TABLE 7

CRUDE	
TIA JUANA LIGHT	SL.121C-AN.69

HYDROCARBON COMPONENT ANALYSIS

15/5 Cut Point	°F VT	158-212	Summary:		
15/5 Cut Point	°C VT	70-100	Paraffins	Vol. %	62.6
Yield Cut Range	Vol. %	5.9-9.8	Naphthenes	Vol. %	32.8
Yield On Crude	Vol. %	3.9	Aromatics	Vol. %	4.6

GAS CHROMATOGRAPHY	Vol. % On Naphtha		Vol. % On Naphtha
Methane		2,4-Dimethylhexane	0.2
Ethane		Methylcyclohexane	7.7
Propane		3,3-Dimethylhexane	
Isobutane		2,2,3-Trimethylpentane	
n-Butane		2,3,4-Trimethylpentane	
Isopentane	0.3	2-Methyl-3-Ethylpentane	
n-Pentane	0.3	1-Trans-2-Cis-4-Trimethylcyclopentane	0.2
2,2-Dimethylbutane			
2,3-Dimethylbutane		Ethylcyclopentane	0.6
2-Methylpentane	1.8	1-Trans-2-Cis-3-Trimethylcyclopentane	0.1
3-Methylpentane	1.9		
Cyclopentane		2,3-Dimethylhexane	
n-Hexane	11.1	2-Methylheptane	0.2
2,2-Dimethylpentane	0.3	4-Methylheptane	
2,4-Dimethylpentane	1.0	3,4-Dimethylhexane	
2,2,3-Trimethylbutane		3-Methylheptane	
Methylcyclopentane	5.9	3-Ethylhexane	
3,3-Dimethylpentane	0.2	1-Cis-3-Dimethylcyclohexane	
2-Methylhexane	9.1	1-Trans-4-Dimethylcyclohexane	
2,3-Dimethylpentane	3.6	1,1-Dimethylcyclohexane	
3-Methylhexane	11.1	1-Methyl-Trans-3-Ethylcyclopentane	
Cyclohexane	8.1		
2,2,4-Trimethylpentane		1-Methyl-Cis-3-Ethylcyclopentane	
1,1-Dimethylcyclopentane	1.5		
3-Ethylpentane	1.0	1-Methyl-Trans-2-Ethylcyclopentane	
1-Cis-3-Dimethylcyclopentane	2.3		
1-Trans-3-Dimethylcyclopentane	2.1	n-Octane	
1-Trans-2-Dimethylcyclopentane	3.8	2,2,4-Trimethylhexane	
n-Heptane	20.3	1-Trans-2-Dimethylcyclohexane	
1-Cis-2-Dimethylcyclopentane	0.3	2,4,4-Trimethylhexane	
Benzene	1.7	2,3,5-Trimethylhexane	
2,2-Dimethylhexane		Toluene	2.9
1,1,3-Trimethylcyclopentane	0.2	1-Trans-3-Dimethylcyclohexane	
2,5-Dimethylhexane	0.2	1-Cis-4-Dimethylcyclohexane	

TABLE 8

CRUDE	
TIA JUANA LIGHT	SL.121C-AN.69

HYDROCARBON COMPONENT ANALYSIS

15/5 Cut Point	°F VT	212-302
15/5 Cut Point	°C VT	100-150
Yield Cut Range	Vol. %	9.8-17.7
Yield on Crude	Vol. %	7.9

Summary:		
Paraffins	Vol. %	52.3
Naphthenes	Vol. %	34.3
Aromatics	Vol. %	13.4

<table>
<tr><th>GAS CHROMATOGRAPHY</th><th>Vol. % On Naphtha</th></tr>
<tr><td>n-Hexane</td><td></td></tr>
<tr><td>2,2-Dimethylpentane</td><td></td></tr>
<tr><td>2,4-Dimethylpentane</td><td></td></tr>
<tr><td>2,2,3-Trimethylbutane</td><td></td></tr>
<tr><td>Methylcyclopentane</td><td></td></tr>
<tr><td>3,3-Dimethylpentane</td><td></td></tr>
<tr><td>2-Methylhexane</td><td>0.4</td></tr>
<tr><td>2,3-Dimethylpentane</td><td>0.1</td></tr>
<tr><td>3-Methylhexane</td><td>0.7</td></tr>
<tr><td>Cyclohexane</td><td>0.1</td></tr>
<tr><td>2,2,4-Trimethylpentane</td><td></td></tr>
<tr><td>1,1,2-Trimethylcyclopentane</td><td>0.1</td></tr>
<tr><td>3-Ethylpentane</td><td></td></tr>
<tr><td>1-Cis-3-Dimethylcyclopentane</td><td>0.1</td></tr>
<tr><td>1-Trans-3-Dimethylcyclopentane</td><td>0.1</td></tr>
<tr><td>1-Trans-2-Dimethylcyclopentane</td><td>0.3</td></tr>
<tr><td>n-Heptane</td><td>5.4</td></tr>
<tr><td>1-Cis-2-Dimethylcyclopentane</td><td></td></tr>
<tr><td>Benzene</td><td></td></tr>
<tr><td>2,2-Dimethylhexane</td><td></td></tr>
<tr><td>1,1,3-Trimethylcyclopentane</td><td>0.4</td></tr>
<tr><td>2,5-Dimethylhexane</td><td>0.5</td></tr>
<tr><td>2,4-Dimethylhexane</td><td>0.7</td></tr>
<tr><td>Methylcyclohexane</td><td>5.6</td></tr>
<tr><td>3,3-Dimethylhexane</td><td>0.1</td></tr>
<tr><td>2,2,3-Trimethylpentane</td><td></td></tr>
<tr><td>2,3,4-Trimethylpentane</td><td></td></tr>
<tr><td>2 Methyl 3 Ethyl Pentane</td><td>0.2</td></tr>
<tr><td>1-Trans-2-Cis-4-Trimethylcyclopentane</td><td>0.6</td></tr>
<tr><td>Ethylcyclopentane</td><td>0.7</td></tr>
<tr><td>1-Trans-2-Cis-3-Trimethylcyclopentane</td><td>0.7</td></tr>
<tr><td>2,3-Dimethylhexane</td><td>0.6</td></tr>
<tr><td>2-Methylheptane</td><td>6.2</td></tr>
<tr><td>4-Methylheptane</td><td>2.2</td></tr>
<tr><td>3,4-Dimethylhexane</td><td>0.6</td></tr>
<tr><td>3-Methylheptane</td><td>4.1</td></tr>
<tr><td>3-Ethylhexane</td><td>0.1</td></tr>
<tr><td>1-Cis-3-Dimethylcyclohexane</td><td rowspan="3">4.3</td></tr>
<tr><td>1-Trans-4-Dimethylcyclohexane</td></tr>
<tr><td>1,1-Dimethylcyclohexane</td></tr>
<tr><td>1-Methyl-Trans-3-Ethylcyclopentane</td><td rowspan="3">0.8</td></tr>
<tr><td>1-Methyl-Cis-3-Ethylcyclopentane</td></tr>
<tr><td>1-Methyl-Trans-2-Ethylcyclopentane</td></tr>
<tr><td>n-Octane</td><td>14.1</td></tr>
</table>

<table>
<tr><th></th><th>Vol. % On Naphtha</th></tr>
<tr><td>2,2,4-Trimethylhexane</td><td></td></tr>
<tr><td>1-Trans-2-Dimethylcyclohexane</td><td></td></tr>
<tr><td>2,4,4-Trimethylhexane</td><td></td></tr>
<tr><td>2,3,5-Trimethylhexane</td><td></td></tr>
<tr><td>Toluene</td><td>3.6</td></tr>
<tr><td>1-Trans-3-Dimethylcyclohexane</td><td rowspan="2">1.7</td></tr>
<tr><td>1-Cis-4-Dimethylcyclohexane</td></tr>
<tr><td>2,2-Dimethylheptane</td><td></td></tr>
<tr><td>2,4-Dimethylheptane</td><td></td></tr>
<tr><td>2,6-Dimethylheptane</td><td></td></tr>
<tr><td>4,4-Dimethylheptane</td><td></td></tr>
<tr><td>1-Methyl-Cis-2-Ethylcyclopentane</td><td></td></tr>
<tr><td>2,5-Dimethylheptane</td><td></td></tr>
<tr><td>3,5-Dimethylheptane</td><td></td></tr>
<tr><td>3,3-Dimethylheptane</td><td></td></tr>
<tr><td>1-Cis-2-Dimethylcyclohexane</td><td>0.5</td></tr>
<tr><td>Ethylcyclohexane</td><td>4.5</td></tr>
<tr><td>2 Methyl 3 Ethyl Hexane</td><td></td></tr>
<tr><td>2,3,3-Trimethylhexane</td><td></td></tr>
<tr><td>3,3,4-Trimethylhexane</td><td></td></tr>
<tr><td>2,3-Dimethylheptane</td><td></td></tr>
<tr><td>3,4-Dimethylheptane</td><td></td></tr>
<tr><td>4-Ethylheptane</td><td></td></tr>
<tr><td>4-Methyloctane</td><td>1.3</td></tr>
<tr><td>1,1,4-Trimethylcyclohexane</td><td></td></tr>
<tr><td>2-Methyloctane</td><td>3.7</td></tr>
<tr><td>3-Ethylheptane</td><td></td></tr>
<tr><td>3-Methyloctane</td><td>3.2</td></tr>
<tr><td>2,3-Dimethyl-3-Ethylpentane</td><td></td></tr>
<tr><td>3,3-Diethylpentane</td><td></td></tr>
<tr><td>1-Cis-3-Trans-5-Trimethylcyclohexane</td><td></td></tr>
<tr><td>1-Trans-2-Trans-4-Trimethylcyclohexane</td><td></td></tr>
<tr><td>n-Nonane</td><td>7.4</td></tr>
<tr><td>C_9 Naphthenes</td><td></td></tr>
<tr><td>Ethylbenzene</td><td>1.8</td></tr>
<tr><td>Paraxylene</td><td rowspan="2">5.5</td></tr>
<tr><td>Metaxylene</td></tr>
<tr><td>Orthoxylene</td><td>1.8</td></tr>
<tr><td>C_9 Isoparaffins</td><td></td></tr>
<tr><td>n-Decane</td><td></td></tr>
<tr><td>C_{10} Plus Naphthenes</td><td></td></tr>
<tr><td>Isopropylbenzene</td><td rowspan="4">0.7</td></tr>
<tr><td>n-Propylbenzene</td></tr>
<tr><td>Meta-Ethyl-Toluene</td></tr>
<tr><td>Para-Ethyl-Toluene</td></tr>
<tr><td>Other C_9 & Heavier Paraffins</td><td>0.7</td></tr>
<tr><td>C_9 & Heavier Naphthenes</td><td>13.8</td></tr>
</table>

TABLE 9

CRUDE	
TIA JUANA LIGHT	SL.121C-AN.69

HYDROCARBON TYPE ANALYSIS

15/5 Cut Point	°F VT	212-248	248-275	275-302	302-320
Range of Cut	Vol. %	9.8-12.9	12.9-15.2	15.2-17.7	17.7-19.4
Yield on Crude	Vol. %	3.1	2.3	2.5	1.7
Yield on Cut:					
Paraffins	Vol. %	52.3	52.6	50.7	52.8
Naphthenes	Vol. %	37.2	34.2	32.5	31.4
Aromatics	Vol.%	10.5	13.2	16.8	15.8

15/5 Cut Point	°F VT	320-347	347-374	374-401	401-428
Range of Cut	Vol. %	19.4-22.0	22.0-24.6	24.6-27.2	27.2-29.8
Yield on Crude	Vol. %	2.6	2.6	2.6	2.6
Yield on Cut:					
Paraffins	Vol. %	46.3	45.2	48.7	42.1
Naphthenes	Vol. %	32.2	33.1	34.8	40.4
Aromatics	Vol. %	21.5	21.7	16.5	17.5

15/5 Cut Point	°F VT	428-455	455-482	482-509	509-536	536-563
Range of Cut	Vol. %	29.8-32.3	32.3-34.8	34.8-37.5	37.5-40.0	40.0-42.
Yield on Crude	Vol. %	2.5	2.5	2.7	2.5	2.6
Yield on Cut:						
Paraffins	Vol. %	48.7	46.4	44.6	43.9	41.3
Naphthenes	Vol. %	35.8	34.1	32.8	29.6	28.1
Aromatics	Vol. %	15.5	19.5	22.6	26.5	30.6

TABLE 10

CRUDE	
TIA JUANA LIGHT	SL.121C-AN.69

KEROSENE & TURBO FUELS

15/5 Cut Point	°F VT	302-401	302-455	302-536	374-536	401-509
15/5 Cut Point	°C VT	150-205	150-235	150-280	190-280	205-265
Yield Cut Range	Vol. %	17.7-27.2	17.7-32.3	17.7-40.0	24.6-40.0	27.2-37.5
Yield On Crude	Vol. %	9.5	14.6	22.3	15.4	10.3
Mid-Point	Vol. %	22.5	25.0	28.9	32.3	32.4
Gravity	°API	48.2	46.2	43.2	40.7	40.6
Specific Gravity	60/60	0.7874	0.7963	0.8100	0.8217	0.8222
Total Sulfur	Wt. %	0.021	0.045	0.14	0.19	0.17
Mercaptan Sulfur	ppm	2	1	1	1	2
Smoke Point	mm	24	23	22	20	20
Lum. No.		53.0	49.9	45.5	41.8	41.7
Freezing Point, IP	°F	-89	-76	-54	-38	-37
Cloud Point	°F	> -80	-78	-56	-40	-38
Pour Point	°F	> -90	-80	-60	-45	-40
Aniline Point	°F	133	136	143	150	150
Diesel Index		64	63	62	61	61
Color	Saybolt	+30				+24
Refractive Index @ 67°C		1.4180	1.4238	1.4309	1.4380	1.4383
Paraffins	Vol. %	47.8	46.9	46.2	45.7	45.4
Naphthenes	Vol. %	33.0	34.8	33.9	34.6	35.8
Aromatics	Vol. %	19.2	18.3	19.9	19.7	18.8
Aromatics, FIA	Vol. %	18.7				18.0
Viscosities:						
Kinematic @ –30°F	cs	3.91	5.30	9.30	16.0	16.2
@ 0°F		2.53	3.35	5.10	7.70	7.80
@ 100°F		1.00	1.17	1.48	1.87	1.88

TABLE 11

CRUDE: TIA JUANA LIGHT — SL.121C-AN.69

MIDDLE DISTILLATES

15/5 Cut Point	°F VT	302-401	320-650	401-509	401-650	509-590	590-698	302-698
15/5 Cut Point	°C VT	150-205	160-343	205-265	205-343	265-310	310-370	150-370
Yield Cut Range	Vol. %	17.7-27.2	19.4-51.1	27.2-37.5	27.2-51.1	37.5-45.2	45.2-55.6	17.7-55.6
Yield On Crude	Vol. %	9.5	31.7	10.3	23.9	7.7	10.4	37.9
Mid-Point	Vol. %	22.5	35.3	32.4	39.2	41.4	50.4	36.7
Gravity	°API	48.2	39.3	40.6	36.8	35.5	30.7	38.5
Specific Gravity	60/60	0.7874	0.8285	0.8222	0.8408	0.8473	0.8724	0.8324
Total Sulfur	Wt. %	0.021	0.35	0.17	0.44	0.51	0.91	0.42
Aniline Point	°F	133	154	150	159	162	172	154
Diesel Index		64	61	61	59	58	53	59
Cetane Index			52	49	54	55	58	53
Cloud Point	°F	> -80	-18	-38	-8	2	36	8
Pour Point	°F	> -90	-20	-40	-10	0	35	5
Refractive Index @ 67°C		1.4180	1.4435	1.4383	1.4498	1.4533	1.4672	1.4457
Neut. No.	mg KOH/gm		0.08		0.09			0.09
Viscosities:								
Kinematic @ 100°F	cs	1.00	2.30	1.88	3.08	3.70	8.10	2.55
@ 122°F		0.87	1.83	1.54	2.40	2.82	5.70	2.02
@ 150°F		0.74	1.46	1.23	1.83	2.12	4.05	1.57
@ 210°F		0.54	0.97	0.85	1.17	1.32	2.17	1.03
Saybolt Universal @ 100°F	sec.		33.6	32.2	36.2	38.2	52.4	34.5
@ 122°F			32.0		34.0	35.4	44.7	32.7
A. R.				7.6	8.8	9.0	9.9	
N. R.				25.5	26.3	26.9	29.7	

CRUDE

TIA JUANA LIGHT

SL.121C-AN.69

GAS OILS

15/5 Cut Point	°F VT	650-752	752-851	851-950	950-1049	650-851	650-1049	851-1049
15/5 Cut Point	°C VT	343-400	400-455	455-510	510-565	343-455	343-565	455-565
Yield Cut Range	Vol. %	51.1-60.3	60.3-68.8	68.8-76.2	76.2-82.1	51.1-68.8	51.1-82.1	68.8-82.1
Yield On Crude	Vol. %	9.2	8.5	7.4	5.9	17.7	31.0	13.3
Mid-Point	Vol. %	55.7	64.6	72.5	79.2	60.0	66.6	75.5
Gravity	°API	27.9	24.7	21.9	19.3	26.3	23.9	20.7
Specific Gravity	60/60	0.8877	0.9059	0.9224	0.9383	0.8964	0.9106	0.9297
Total Sulfur	Wt. %	1.08	1.20	1.35	1.55	1.14	1.28	1.44
Aniline Point	°F	178	189	199	207	183	192	202
Con Carbon	Wt. %	0.010	0.026	0.40	2.84	0.018	0.66	1.50
Pour Point	°F	55	85	105	120+	70	90	115
Refractive Index @67°C		1.4748	1.4870	1.4975	1.5064	1.4807	1.4896	1.5016
Neut. No.	mg KOH/gm	0.21				0.24		
Nitrogen	Wt. %	0.080	0.102	0.152	0.285	0.091	0.144	0.212
Viscosities:								
Kinematic @ 100°F	cs	14.3	47.5			24.7	68.0	
@ 150°F		6.3	15.0	38.5	108	9.4	20.2	60.0
@ 175°F		4.4	9.5	22.2	55.8	6.35	12.8	33.0
@ 210°F		3.05	5.95	12.3	27.0	4.15	7.65	17.2
Saybolt Universal @ 100°F	sec	74.7	221			118	315	
@ 150°F		46.6	77.6	180	502	57.0	99.0	280
@ 175°F		40.7	57.4	108	260	46.9	69.3	156
@ 210°F		36.4	45.7	67.5	129	39.9	51.2	86.8
Metals:								
Vanadium	ppm							0.68
Nickel	ppm							0.38
Iron	ppm							0.60
A. R.		10.0	11.2	12.0	12.0	10.6	11.0	12.0
N. R.		32.0	34.2	37.6	42.7	32.9	34.8	39.7

TABLE 13

CRUDE	
TIA JUANA LIGHT	SL.121C-AN.69

LUBE DISTILLATES

		WAXY LUBE	DEWAXED LUBE
15/5 Cut Point	°F VT	796-1015	796-1015
15/5 Cut Point	°C VT	424-546	424-546
Yield Cut Range	Vol. %	64.4-80.1	—
Yield On Crude	Vol. %	15.7	14.1
Mid-Point	Vol. %	72.3	—
Gravity	°API	22.2	20.3
Specific Gravity	60/60	0.9206	0.9321
Total Sulfur	Wt. %	1.36	
Con Carbon	Wt. %	0.54	
Aniline Point	°F	198	188
Refractive Index @ 67°C		1.4960	
Neut. No.	mg KOH/gm	0.24	
Nitrogen	Wt. %	0.14	
Pour Point	°F	110	0
Wax Content	Wt. %	9.1	—
Wax Melting Point	°F	113	
Viscosity Index		74	46
Viscosities:			
Kinematic @ 100°F	cs	(152)	245
@ 150°F		38.1	50.7
@ 175°F		22.6	28.2
@ 210°F		12.4	14.7
Saybolt Universal @ 100°F	sec	704	1137
@ 210°F		67.9	76.9

Phenol Treating Susceptibility*

Phenol Treating Characteristics of Dewaxed Lube Cut

Phenol/Oil Ratio

Treat	VI
Raw Stock	46
1/1	86
2/1	95
3/1	99
Viscosity Gravity Constant	0.871

*From Established Correlations

TABLE 14

CRUDE: TIA JUANA LIGHT | SL.121C-AN.69

RESIDUA

15/5 Cut Point	°F VT	650+	700+	752+	851+	950+	1049+	428+	563+
15/5 Cut Point	°C VT	343+	371+	400+	455+	510+	565+	220+	295+
Yield On Crude	Vol. %	48.9	44.4	39.7	31.2	23.8	17.9	70.2	57.4
Gravity	°API	17.3	16.3	15.1	12.6	9.9	7.1	22.5	19.4
Specific Gravity	60/60	0.9509	0.9574	0.9652	0.9820	1.007	1.0209	0.9188	0.9377
Total Sulfur	Wt. %	1.78	1.84	1.93	2.12	2.35	2.59	1.42	1.64
Con Carbon	Wt. %	9.3	10.2	11.2	13.8	17.2	21.6	6.8	8.1
Nitrogen	Wt. %	0.33	0.36	0.39	0.45	0.52	0.60		
Neut No.	mg KOH/gm	0.46							
Pour Point	°F	45	50	60	75	95	120	15	30
Viscosities:									
Kinematic @ 100°F	cs	890	1590	3100				83.0	315
@ 122°F		365	620	1120				46.2	148
@ 150°F		148	235	400	1680			25.5	67.5
@ 210°F		35.0	50.0	77.0	220	1010	7959	9.6	19.6
Furol @ 122°F	sec	172	292	528					70.6
@ 210°F			25.2	37.6	106	484	(3760)		
@ 275°F						77	320		
Universal @ 210°F	sec	165	234	359	1025			57.8	96.8
Redwood 1 @ 100°F	sec	3600						333	1260
Engler @ 50°C	deg.	46.5						5.95	19.2
Metals:									
Vanadium	ppm	185					450		
Nickel	ppm	25					64		
Iron	ppm	28					48		
MNI	Wt. %	6.4	7.0	7.9	9.7	12.2	14.8	4.2	5.4
VBN		20.7	18.9	16.7	12.4	7.7	2.9	28.7	24.0

TABLE 15

CRUDE	
TIA JUANA LIGHT	SL.121C-AN.69

STRAIGHT REDUCED ASPHALTS

Inspection:	Yield Vol. % On Crude	15/5 Cut Point °F VT	Flash – Cleveland Open Cup	Soft Point °F	Penetration:		Viscosity Furol @ 275°F	Specific Gravity 60/60	App. Rating	Pen. Ratio
					@ 77°F	@ 39.2°F				
SAMPLES INSPECTED										
No. 1	16.6	1075	715	138	36	15	376	1.0231		41.7
No. 2	17.9	1049		135	42	16	320	1.0209		38.1
No. 3	19.4	1022		126	65	23	217	1.0151		35.4
No. 4	20.8	995		117	101	32	148	1.0100		31.7
No. 5	23.8	950		100	240	65	80	1.0007		27.1
STANDARD GRADES (From Correlated Data)										
180/200	22.9	965		106	190	54	95	1.0035		28.4
85/100	20.6	1002		120	95	30	160	1.0114		31.6
71/80	19.8	1015		124	75	25	195	1.0136		33.3
51/60	18.8	1032		129	55	20	250	1.0173		36.4
31/40	16.6	1075	715	140	35	14	455	1.0246		40.0

GRAPH NO. 1

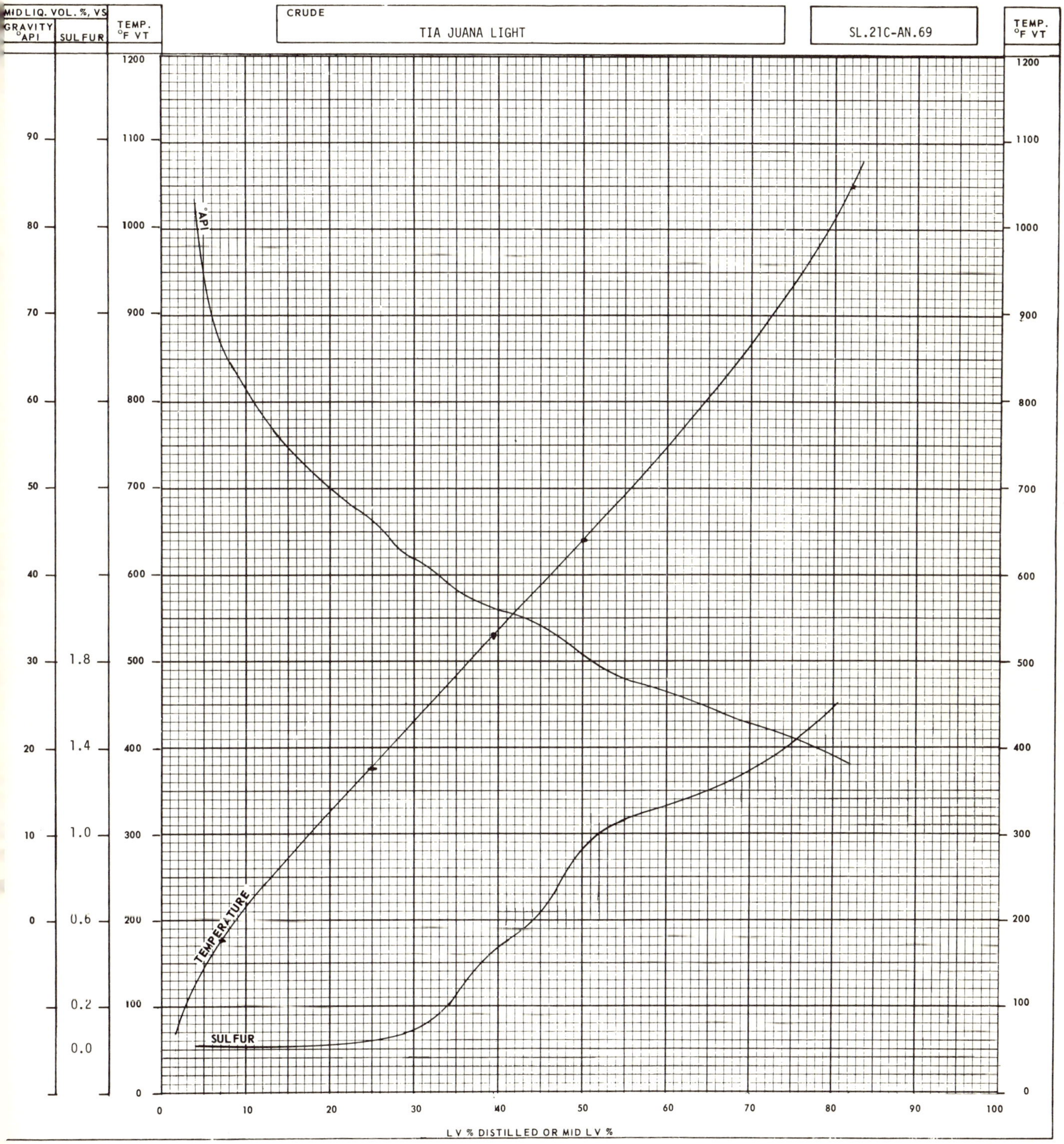
MID LIQ. VOL. %, VS
GRAVITY °API
SULFUR
TEMP. °F VT
CRUDE
TIA JUANA LIGHT
SL.21C-AN.69
TEMP. °F VT
°API
TEMPERATURE
SULFUR
LV % DISTILLED OR MID LV %

GRAPH NO. 2
GASOLINES

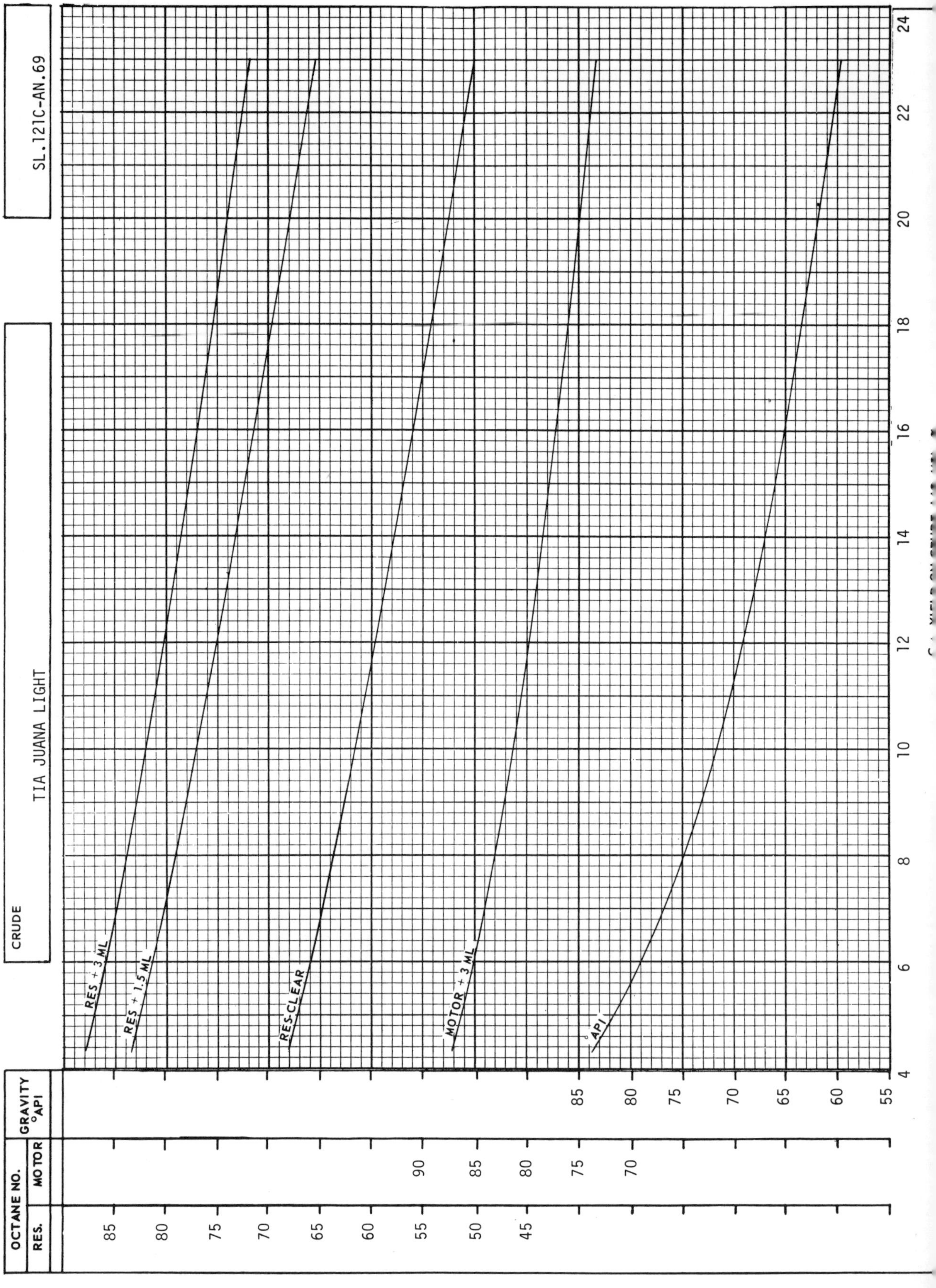

GRAPH NO. 3

GASOLINES

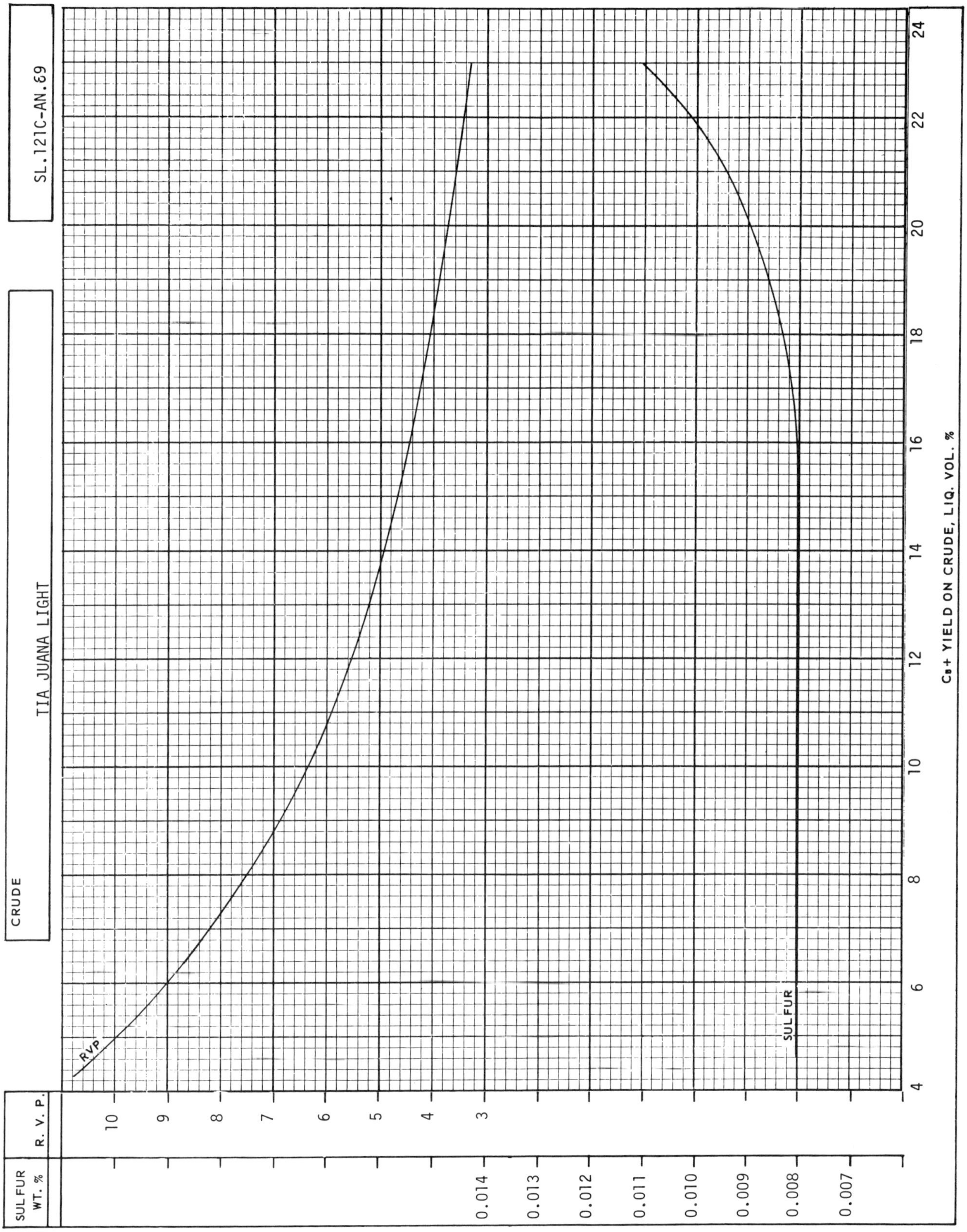

GRAPH NO. 4

NAPHTHAS

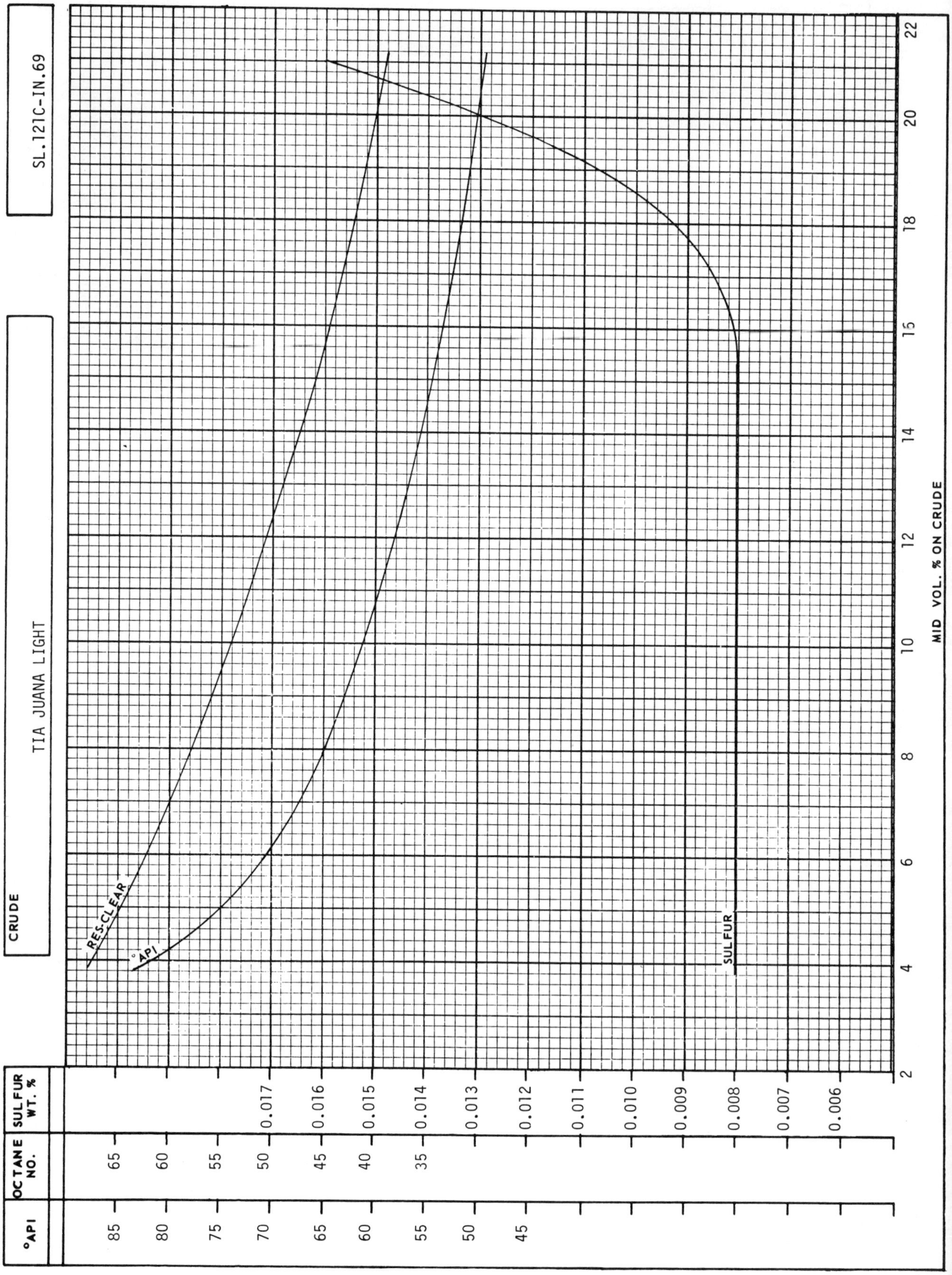

GRAPH NO, 5

KEROSENES

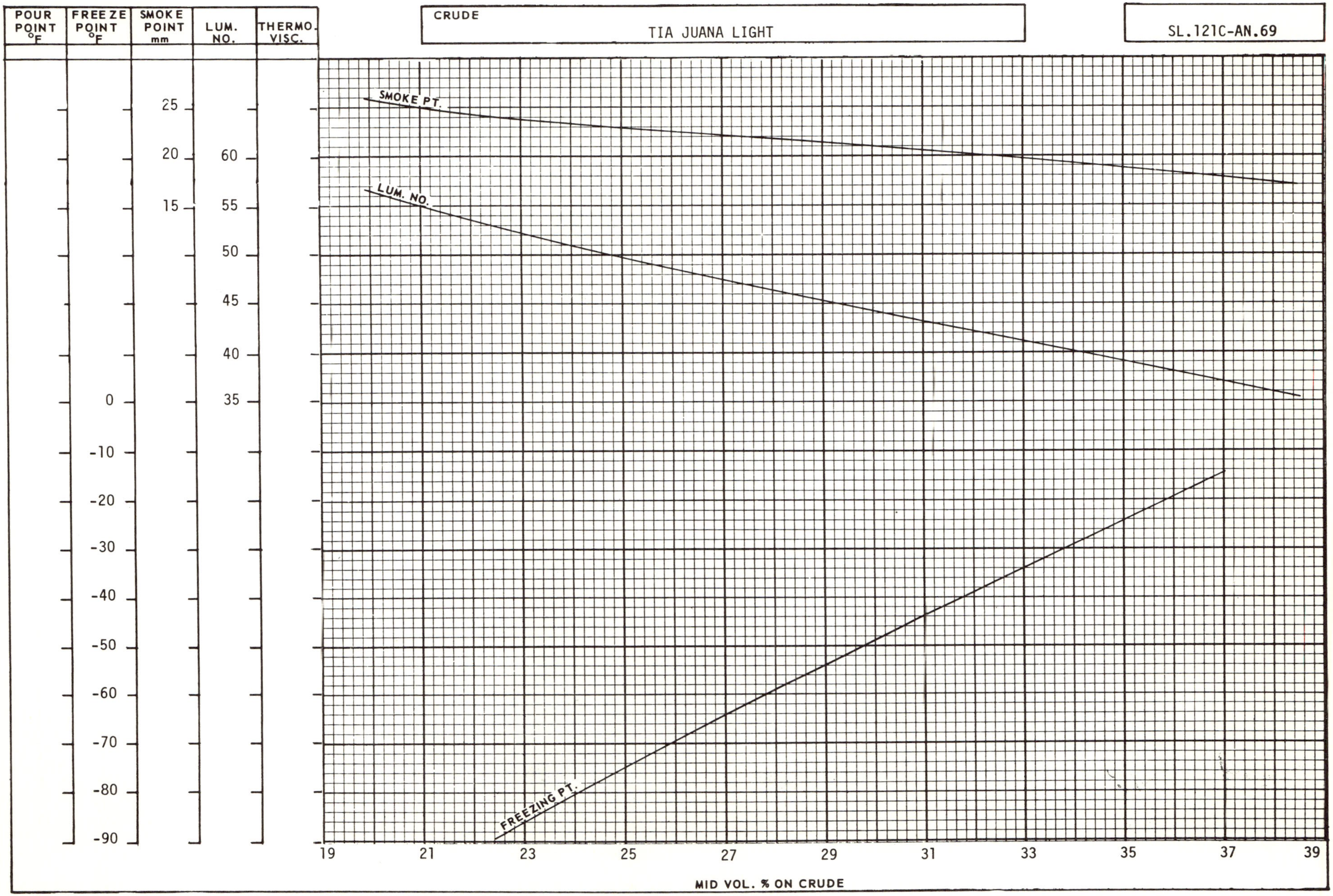

GRAPH NO. 6

KEROSENES

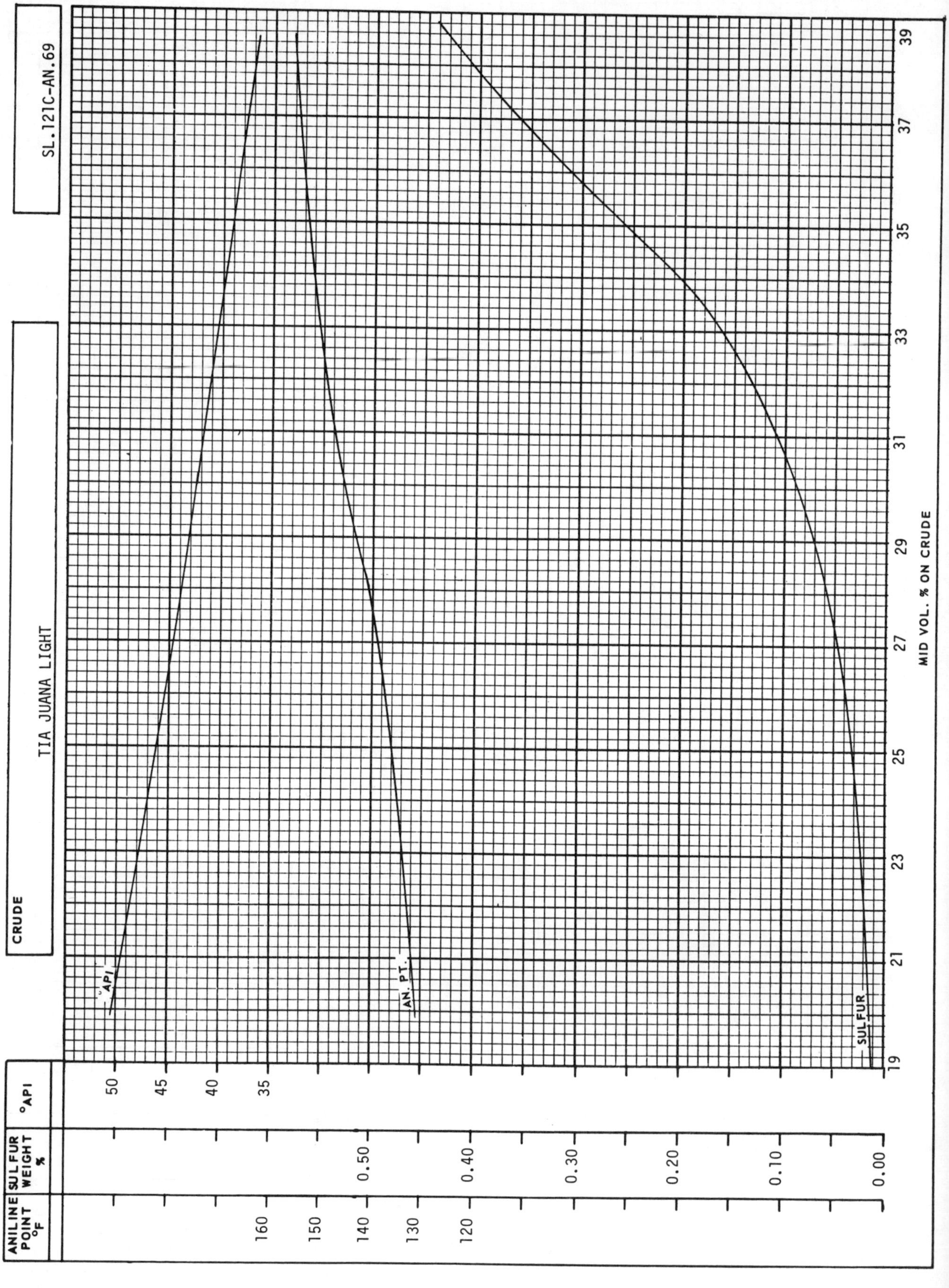

GRAPH NO. 7

MIDDLE DISTILLATES AND GAS OILS

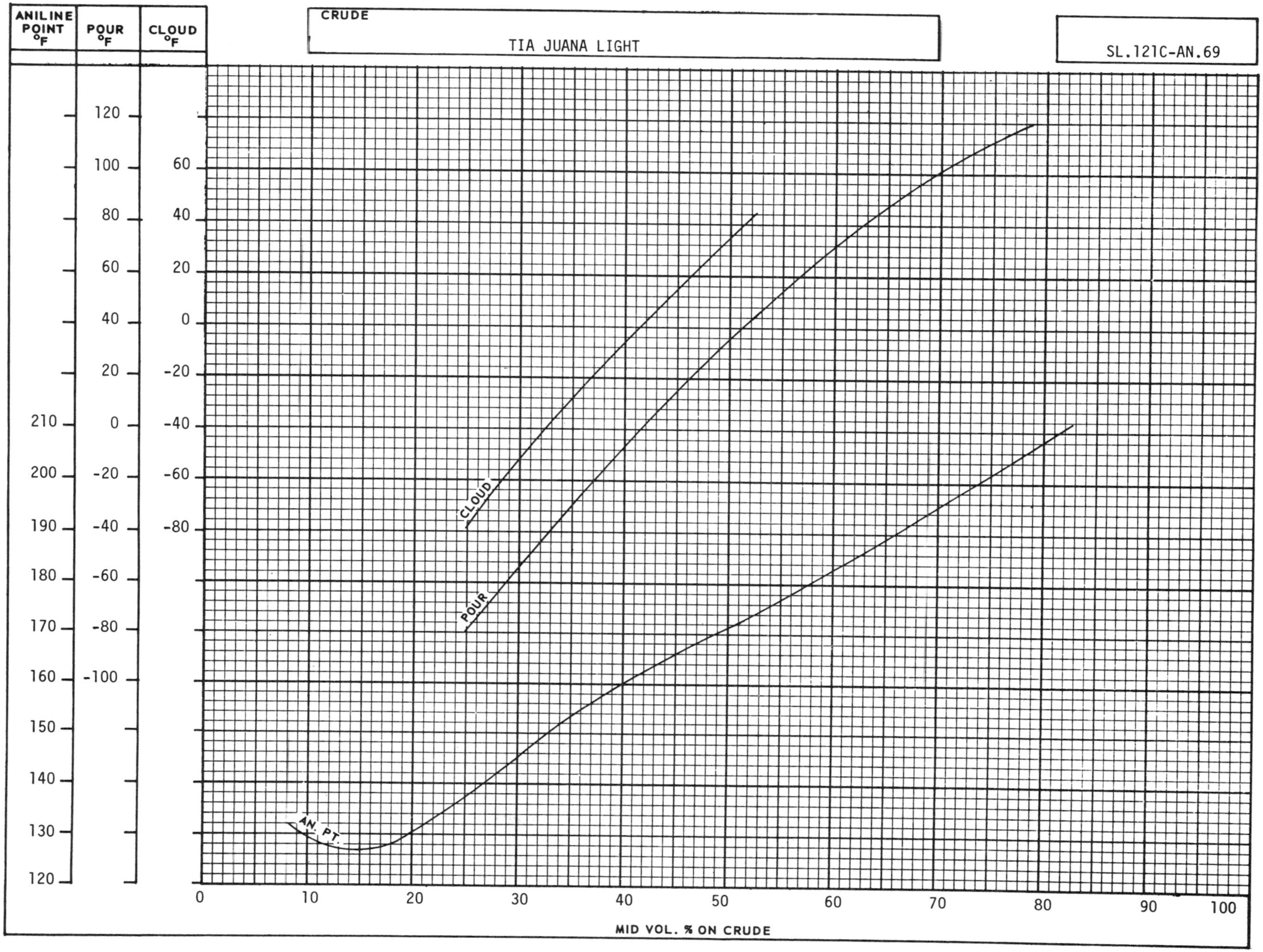

GRAPH NO. 7A

MIDDLE DISTILLATES AND GAS OILS

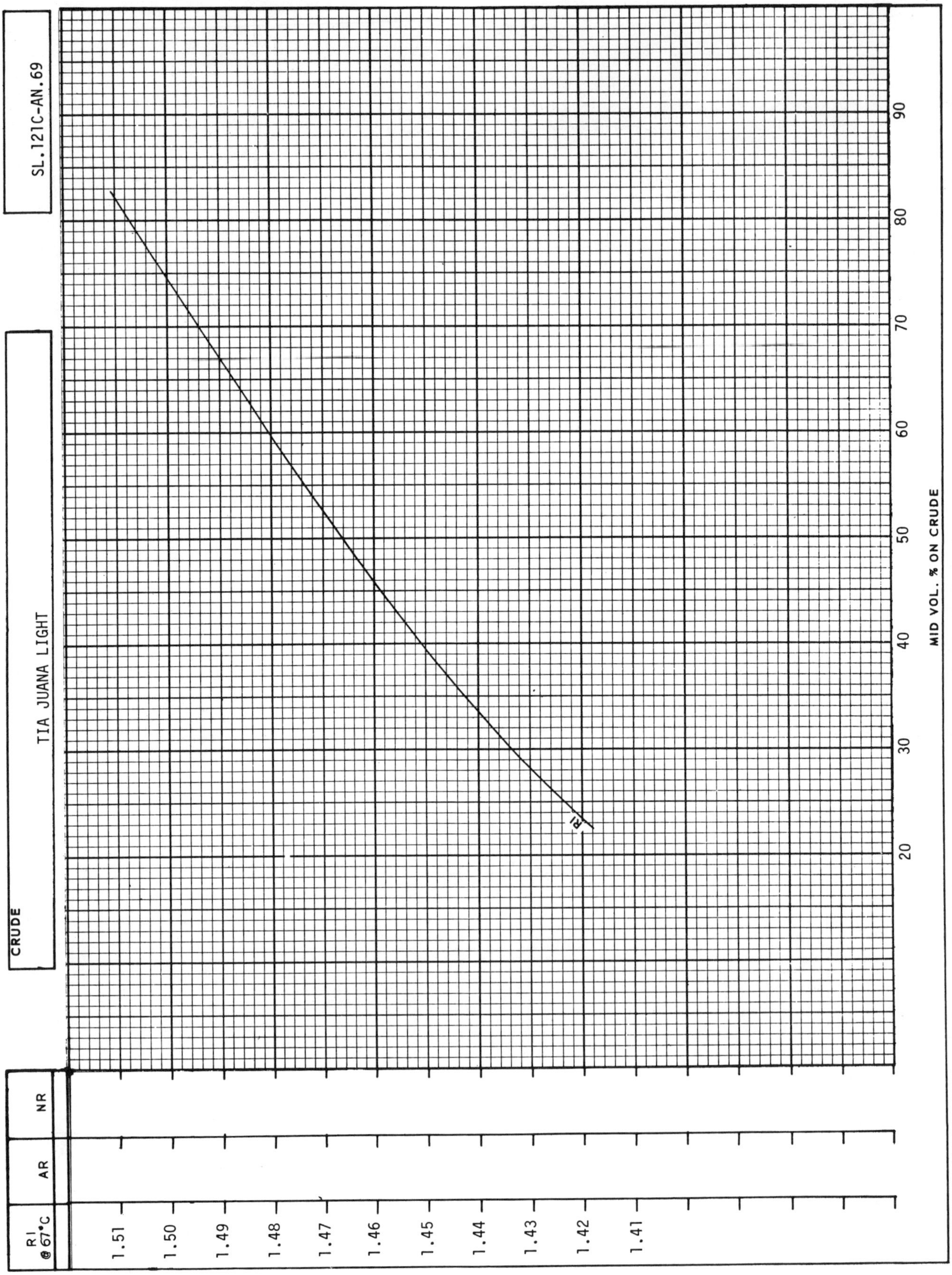

GRAPH NO. 8

MIDDLE DISTILLATES AND GAS OILS

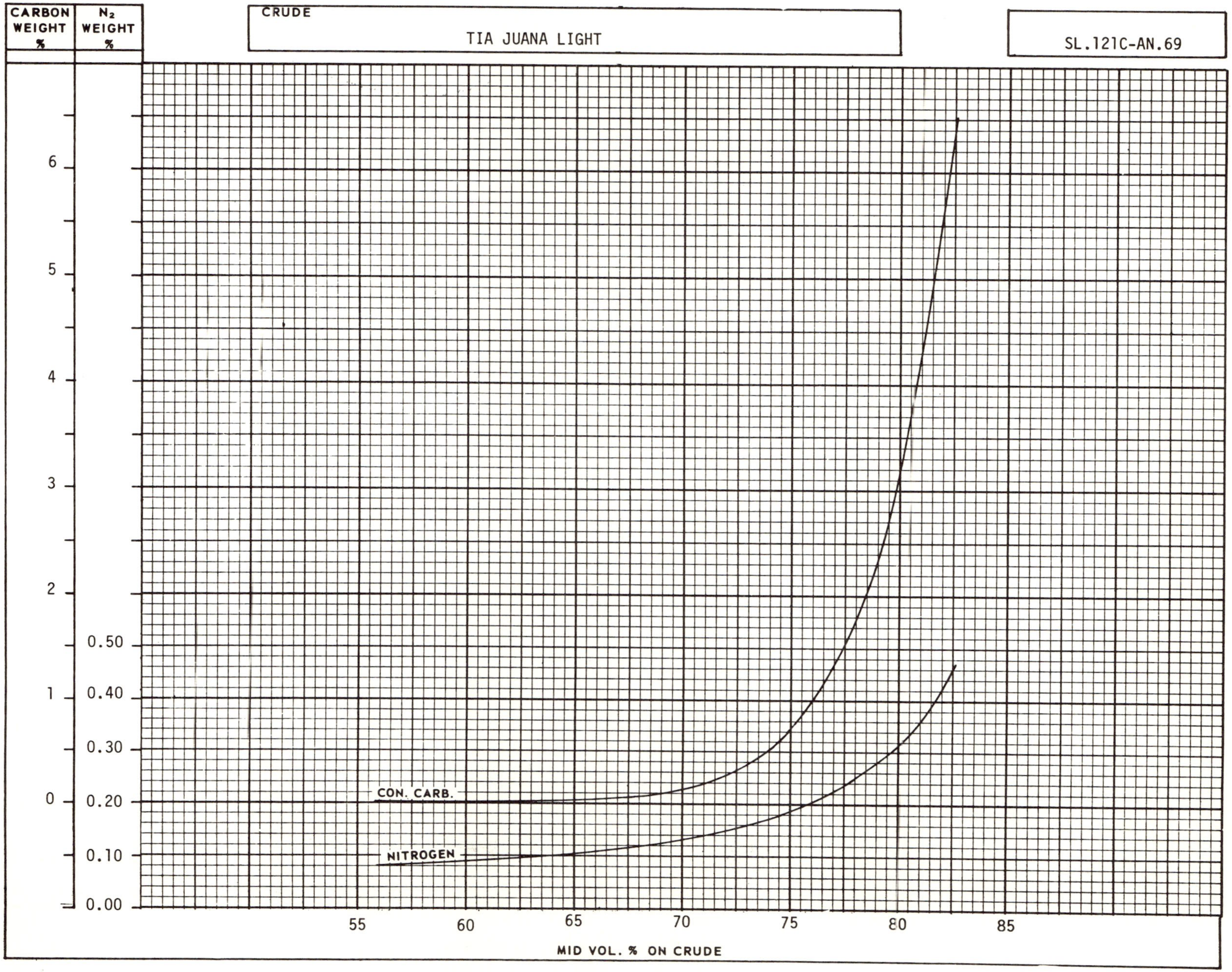

GRAPH NO. 9

MIDDLE DISTILLATES AND GAS OILS

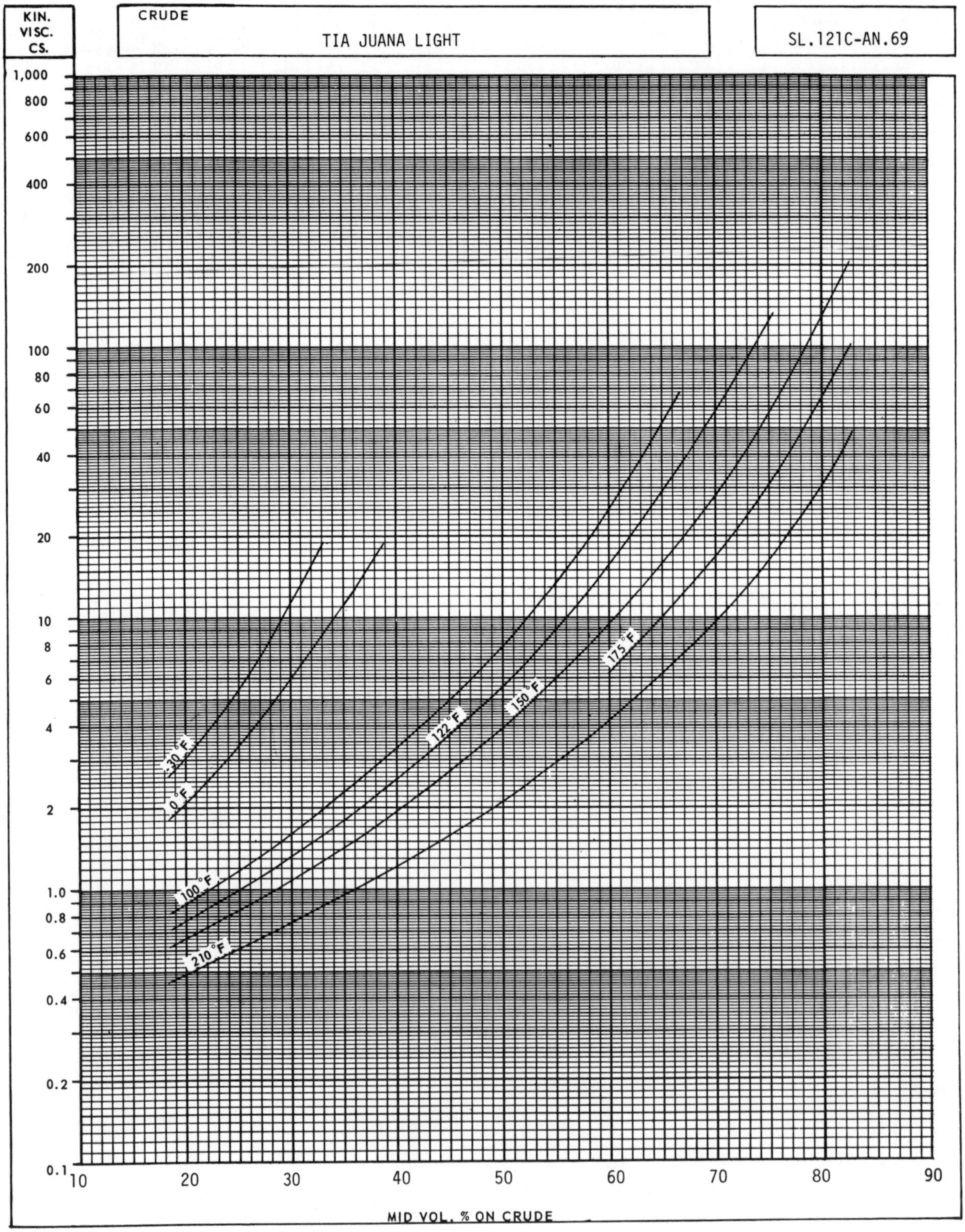

GRAPH NO. 10

MIDDLE DISTILLATES AND GAS OILS

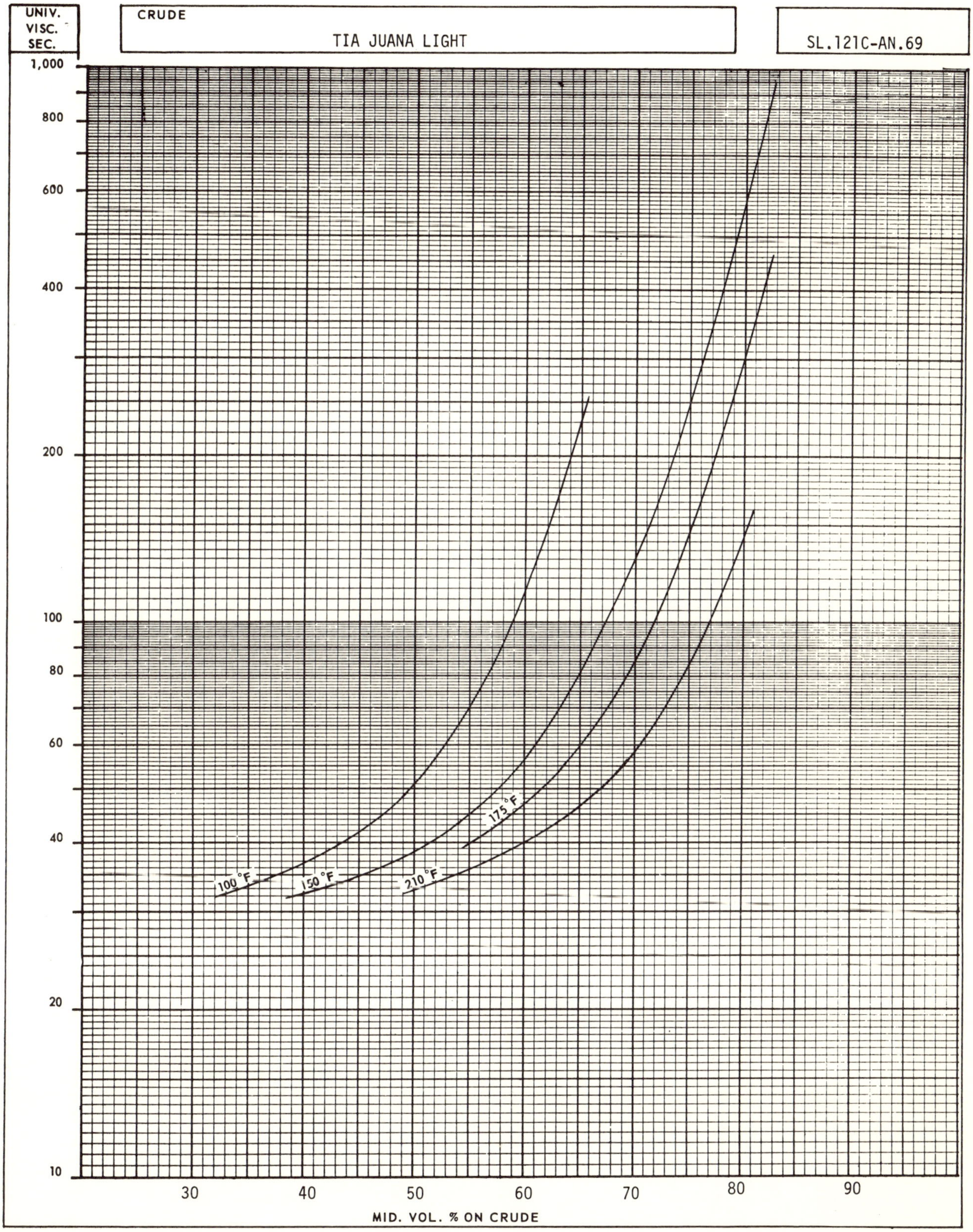

GRAPH NO. 11

RESIDUA

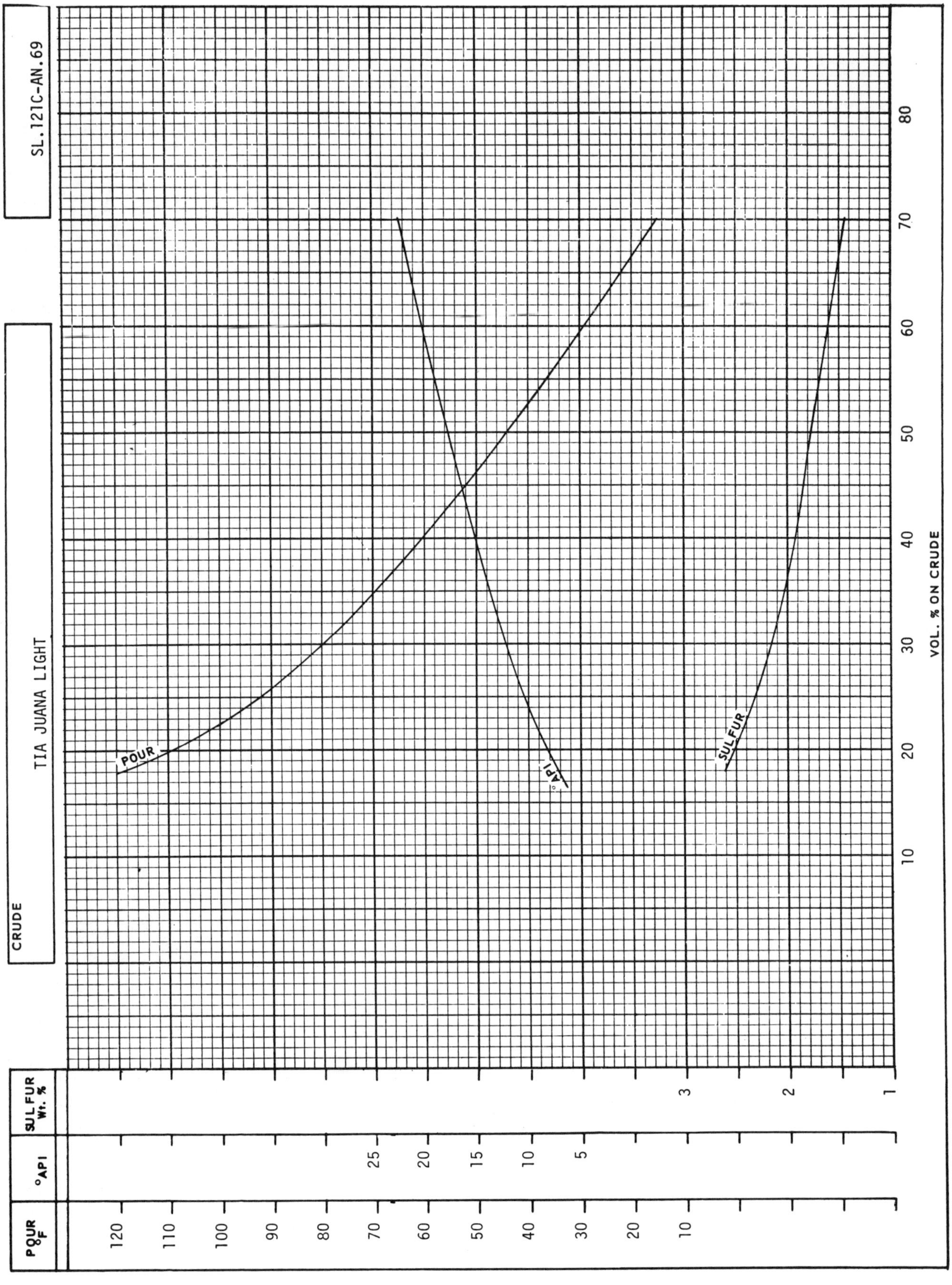

GRAPH NO. 12

RESIDUA

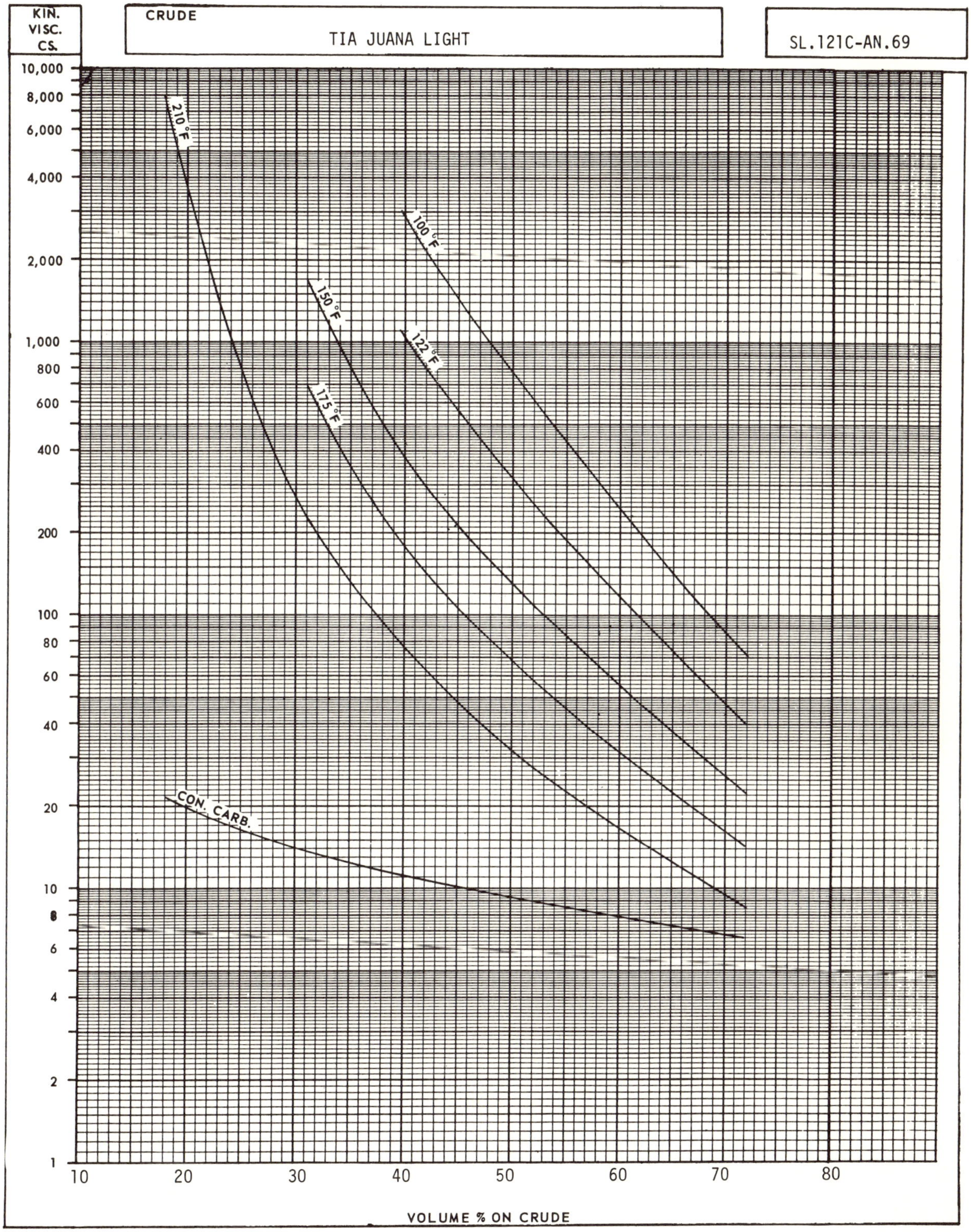

GRAPH NO, 14

RESIDUA

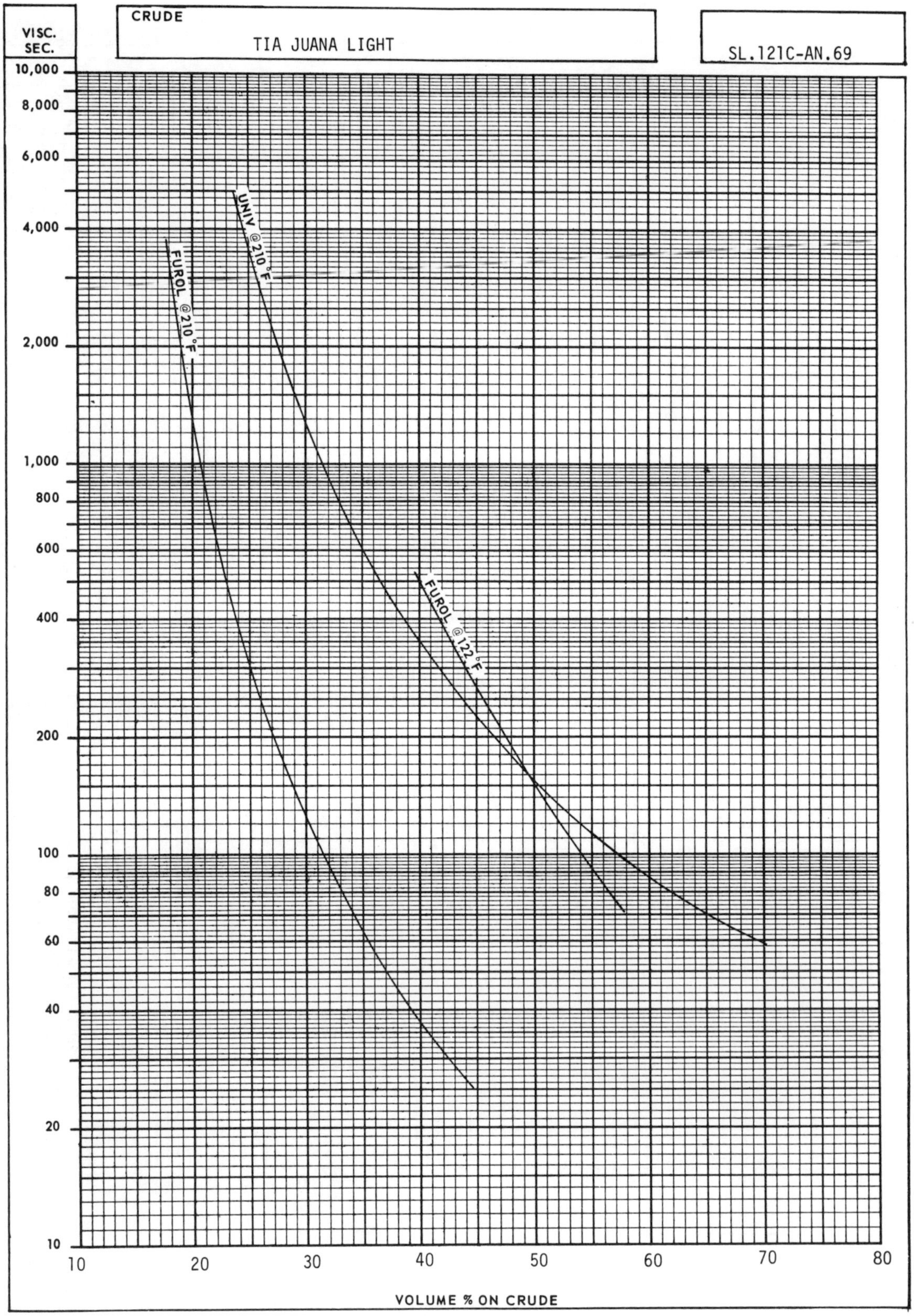

GRAPH NO. 15

STRAIGHT REDUCED ASPHALTS

CRUDE TIA JUANA LIGHT	SL.121C-AN.69

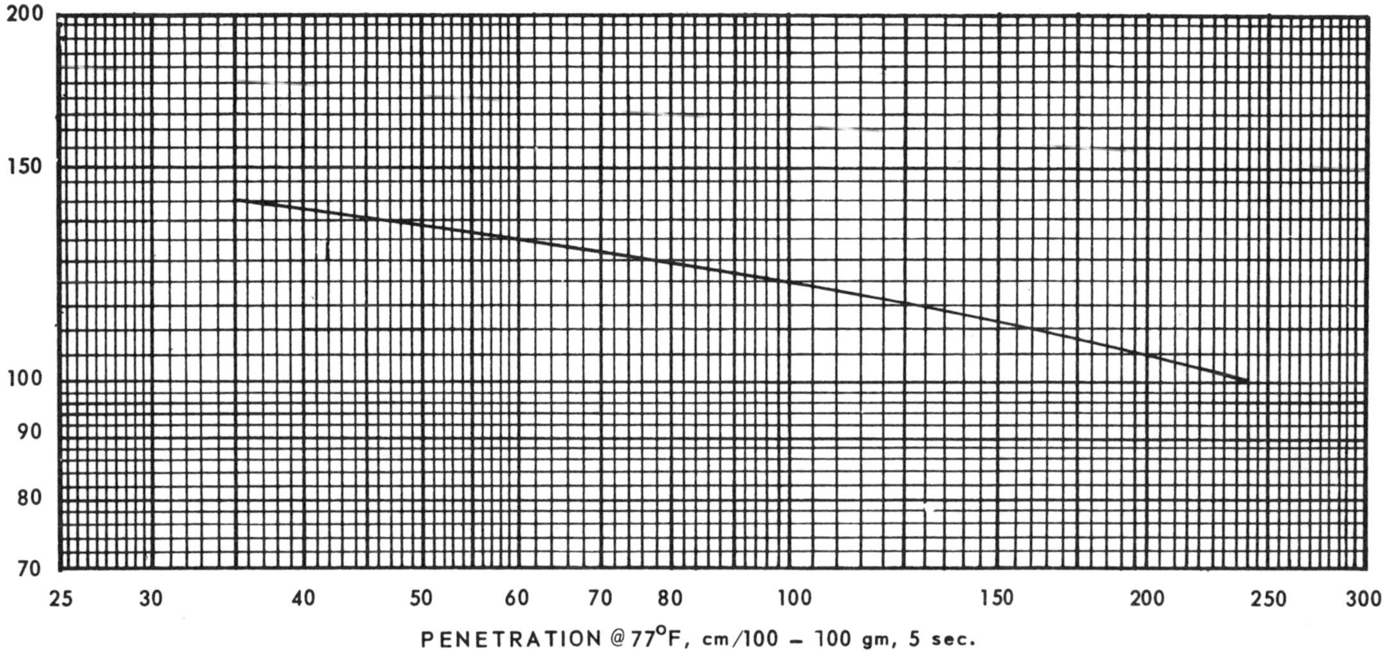

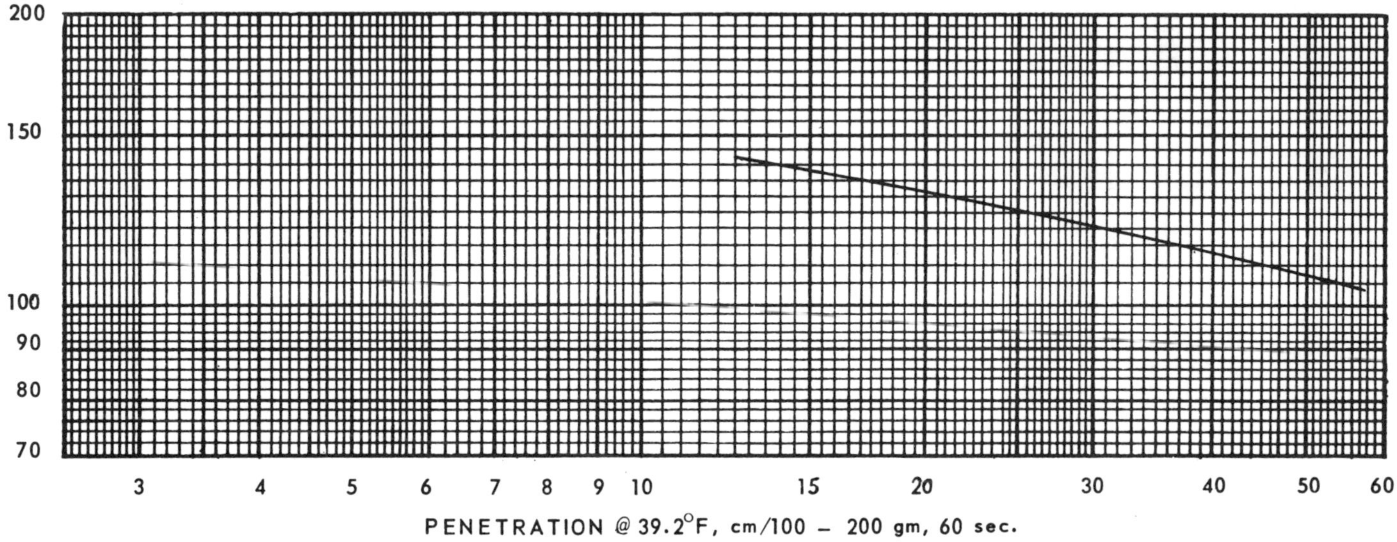

GRAPH NO. 16

STRAIGHT REDUCED ASPHALT

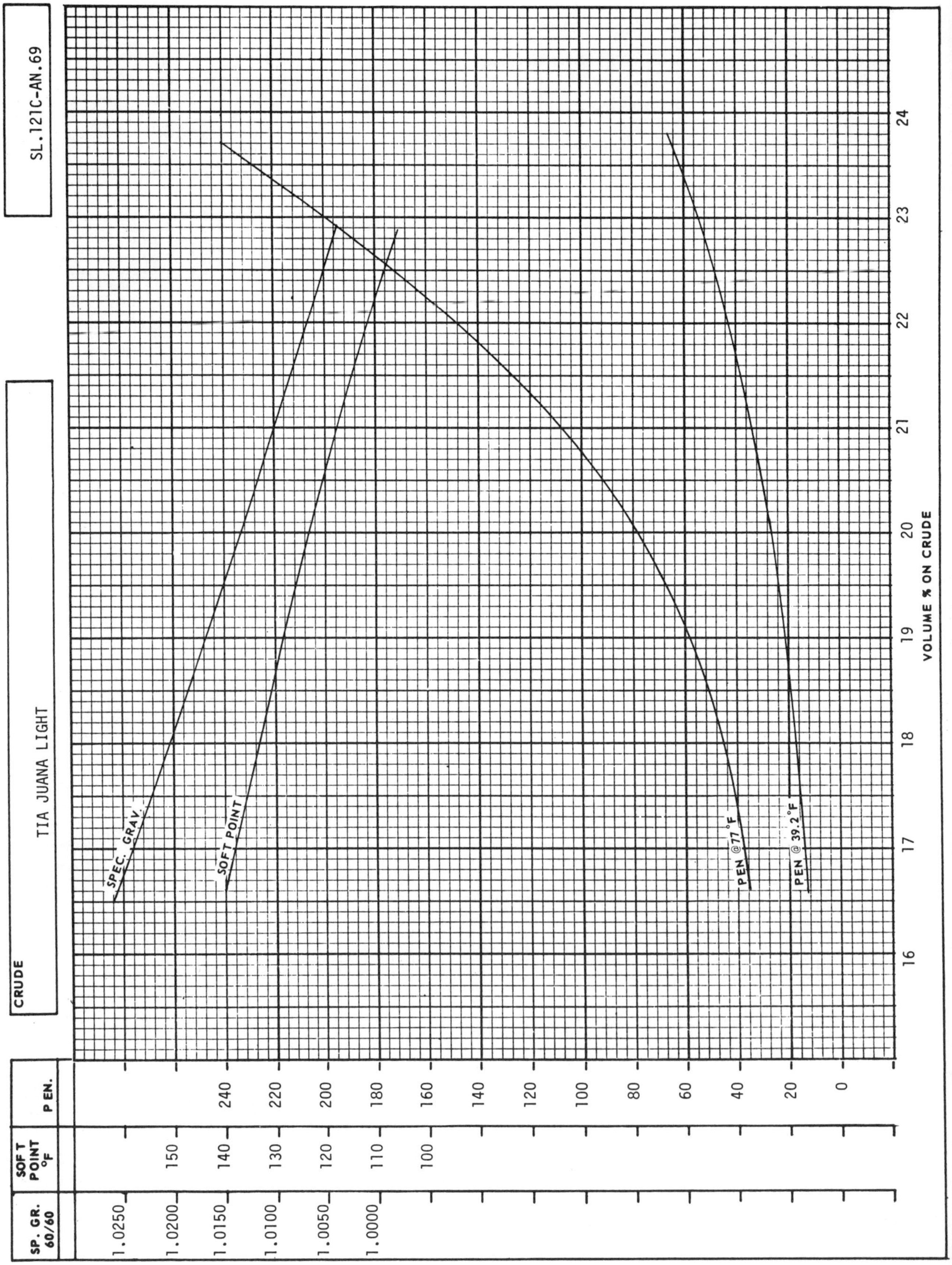

GRAPH NO. 17

STRAIGHT REDUCED ASPHALT VISCOSITY

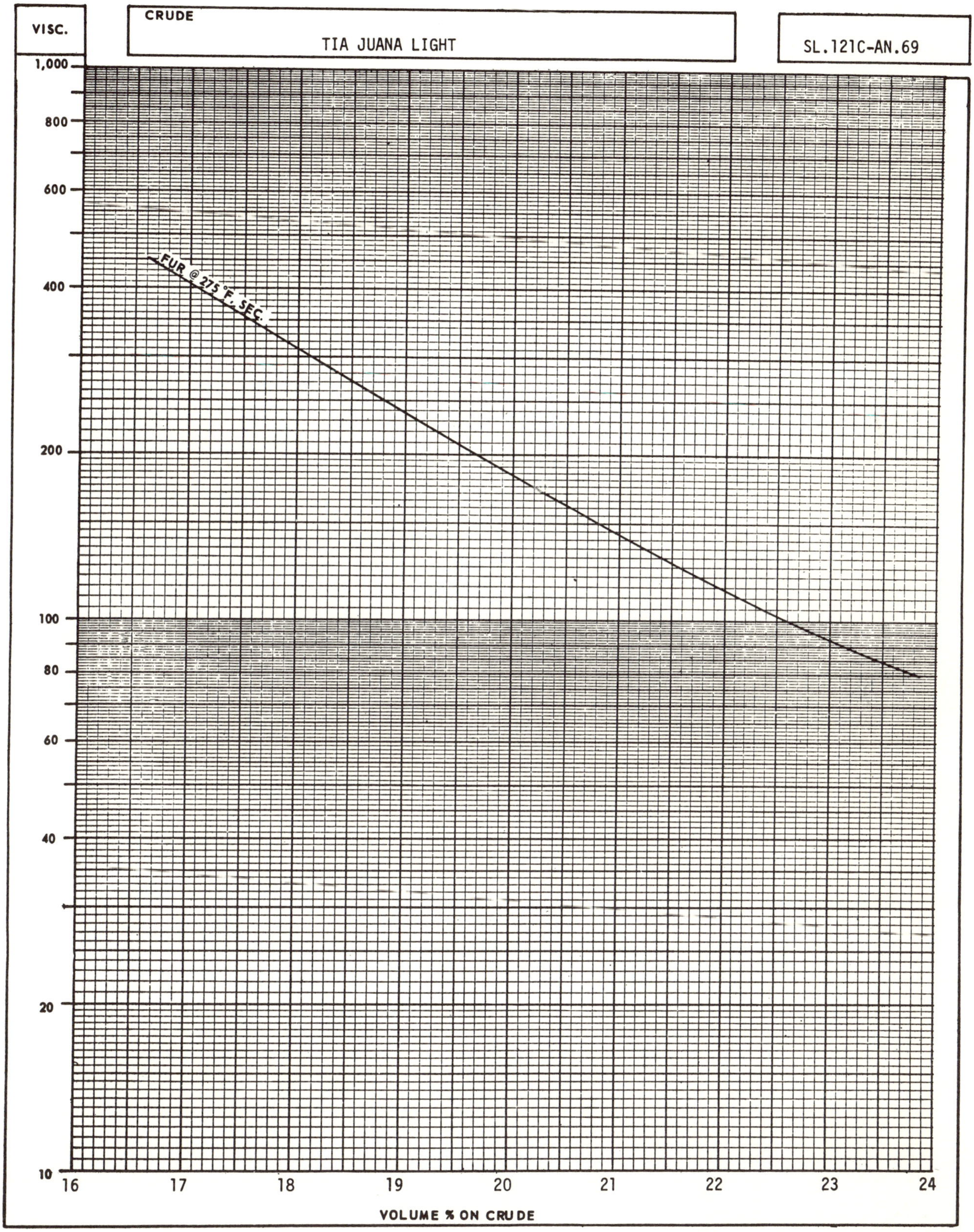

index